Defence Industrial Strategy
Defence White Paper

Presented to Parliament by
The Secretary of State for Defence
By Command of Her Majesty

December 2005

Cm 6697

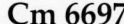

Foreword

The country is rightly proud of its Armed Forces, and recognizes the vital contribution they make to ensuring our security in an uncertain world. Since the Strategic Defence Review in 1998, and through successive White Papers, we have been transforming the Royal Navy, Army and Royal Air Force to face the demands of the 21st Century. Our plans provide for agile, flexible forces that can respond effectively to the varied challenges and opportunities we face now and in the future.

That realignment is supported by a substantial equipment programme, reconfiguring our forces to project power round the world to protect UK interests and strengthen international peace and stability. This has been supported by sustained real increases in the Defence budget arising from each Spending Review since the Government was elected in 1997.

Our Defence Industrial Strategy takes forward our Defence Industrial Policy, published in 2002, by providing greater transparency of our future defence requirements and, for the first time, setting out those industrial capabilities we need in the UK to ensure that we can continue to operate our equipment in the way we choose.

The Defence Industrial Strategy recognises the important contribution that our defence industry makes to delivering military capability and the clarity provided in the strategy will, we believe, promote a dynamic, sustainable and globally competitive defence manufacturing sector. The UK market for defence equipment and services is the second largest in the world, and we recognise that the UK has a broad-based and sophisticated defence industry. The UK offers unique attractions to business and we continue to benefit from the presence in our market of companies, whoever their shareholders are, who are prepared to invest and develop their businesses here.

But Government is clear that there are also challenges ahead. The complex, technologically challenging and high-value systems which we are introducing – and which take many years to design and bring into service – will last for many years. This places increasing emphasis on an ability to support and upgrade them through life, as well as having implications for the level of industrial capability and capacity that it is sensible or economic for industry to retain. We recognize that industry will have to reshape itself, to improve productivity and to adjust to lower production levels once current major equipment projects have been completed, while at the same time retaining the specialist skills and systems engineering capabilities required to manage military capability on a through life basis. Just as the roles and structures of the Armed Forces are transforming, so too are those of the defence industry which has itself to face the future with confidence.

In this Strategy, we consider carefully which industrial capabilities we need to retain in the UK to ensure that we can continue to operate our equipment in the way we choose to maintain appropriate sovereignty and thereby protect our national security. The Strategy sets these out, and explains clearly for the first time which industrial capabilities we require to be sustained onshore, noting that – as now – there are many that we can continue to seek to satisfy through open international competition. In doing so, it builds upon the Defence Industrial Policy, explains more clearly how procurement decisions are made, and to assist industry in planning for the future commits the Government to greater transparency of our forward plans, noting that as in any business, these change over time as spending priorities shift or cost estimates mature.

To implement this strategy will require changes on behalf of both industry and Government. Industry will need to adjust to sustain the capabilities we need once current production peaks are passed. The Government, too, needs to drive forward improvements in the way we acquire, support and upgrade our equipment. Together, the defence industry and government have to change their relationship, working to ensure that our Armed Forces continue to have the equipment they need. Doing this will help ensure the UK defence industry has a sustainable and bright future. This will require continuous effort on both sides over the coming years as it will not be easy. However, by starting the process today, while workloads are high, we can avoid facing a crisis in a few years time. We recognize some companies will find the strategy's conclusions difficult, but believe that industry, the City, the government and the country needs the additional clarity we are offering, to help industry reshape itself for the future.

We will look to the National Defence Industries Council to monitor our joint progress, and will review this Strategy every Spending Review period. In the meantime, and as the basis for the detailed implementation which will follow over the next few months, we commend this Strategy to you.

John Reid
Defence Secretary

Alan Johnson
Trade and Industry Secretary

Des Browne
Chief Secretary to the Treasury

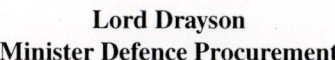

Lord Drayson
Minister Defence Procurement

Alun Michael
Minister of State for Industry
and the Regions

MINISTRY OF DEFENCE

Defence Industrial Strategy Contents

i.   The Defence Industrial Strategy (DIS) is structured in three parts: Part A, providing the strategic context; Part B, reviewing different industrial sectors and cross-cutting industrial capabilities; and Part C, outlining the implications for MOD and industry as a whole, and how the DIS will be implemented.

Part A – Strategic Overview

ii.   The global security environment in which the Armed Forces operate has changed substantially over the past fifteen years. Facing new and complex challenges, the roles, size and shape of Armed Forces have also changed. In parallel, the defence industry has evolved; defence companies are now often transnational, needing to attract and retain investors in international markets – forcing increased efficiency, restructuring and rationalisation. We are now reaching a crossroads.

iii.   Although we are in the middle of a substantial transformation, involving a series of major new platforms (including the future aircraft carriers, Type 45 Destroyers, new medium-weight armoured fighting vehicles, and the A400M, Typhoon and Joint Combat Aircraft), we expect these platforms to have very long service lives. This means the future business for the defence industry in many sectors will be in supporting and upgrading these platforms, rapidly inserting technology to meet emerging threats, fulfil new requirements and respond to innovative opportunities, not immediately moving to design the next generation.

iv.   In parallel, industrial rationalisation continues, and sustaining competition to meet domestic requirements is increasingly difficult. In several sectors, following the entry into service of major projects, there will be substantial overcapacity in production facilities in the UK defence industry in a few years' time.

v.   As we look to non-British sources of supply, whether at the prime or subsystems level, we need to continue to recognise the extent to which this may constrain the choices we can make about how we use our Armed Forces – in other words, how we maintain our sovereignty and national security.

vi.   Companies now have more choice than ever before about which markets to enter, which secure the best return for shareholders, and where to base their operations. If we do not make clear which industrial capabilities we need to have onshore (and this includes those maintained by foreign-owned defence companies), industry will make independent decisions and indigenous capability which is required to maintain our national security may disappear.

vii.   Equally, we do not seek to restrict the scope for international cooperation and competition where this is appropriate, and we cannot afford to maintain a complete cradle-to-grave industrial base in all areas. As industry has told us, greater clarity is therefore needed urgently on which capabilities must be retained onshore, and which by implication can be met from a wider market. The DIS does not seek to set out a preferred route to international restructuring; that is very much industry's business. But it does seek to create a clear UK context to inform these decisions.

Our aim in the DIS

viii.   For these reasons, we need to consider how best the MOD should seek to engage with the industrial base in order to meet our requirements. The DIS flows from the wider Defence Industrial Policy (2002), and is 'driven by the need to provide the Armed Forces with the equipment which they require, on time, and at best value for money for the taxpayer.' The DIS is thus one of many contributions to the wider aim of ensuring that the capability requirements of the Armed Forces can be met, now and in the future.

ix.   The DIS will promote a sustainable industrial base, that retains in the UK those industrial capabilities needed to ensure national security. Our interaction with this industrial base must provide good value to the taxpayer and good returns to shareholders based on delivery of good performance, consistent with broader security and economic policy.

x.   To deliver this, the DIS:

- gives a strategic view of defence capability requirements going forward (including new projects, but also the support and upgrade of equipment already in service), by sector. Part of the strategic view is specifying, in order to meet these, which industrial capabilities we would wish to see retained in the UK for Defence reasons. We aim to communicate the overall view to industry as clearly as possible, recognising that plans change as the strategic or financial environment evolves (and the DIS explains our current internal planning process, to allow industry to make informed judgements about how to interpret this information);

- gives further detail on the principles and processes that underpin procurement and industrial decisions;

- where there is a mismatch between the level of activity our own plans (and export/civil opportunities) would support and that required to sustain desired industrial capabilities onshore, investigates how we might with industry address that gap.

The evolving market and the UK business environment

xi.   We recognise that in the UK we have a successful and sophisticated industrial base with a broad range of capabilities and which delivers a large proportion of our defence equipment and services. We welcome overseas investment where this creates value, employment, technology or intellectual assets in the UK.

xii   We also recognise the attractions of the US market, given its scale and high levels of investment in research and technology, and that the level of influence and attractiveness of MOD business varies by sector and by type of company. But the UK provides a unique environment for the defence industry:

- a greater proportion of our overall business is available to industry than in any other major defence nation, and growing expertise in the combination of systems engineering skills, agility and supply chain management required to deliver through-life capability management gives the UK defence industry a comparative advantage;

- we have a sophisticated demand for high-value products which have to stand up to active service, and consequently, are easier to market to export customers;
- we have an open market and diversity of suppliers which encourages innovation, new entrants and inward investment;
- and profit potential and a trading environment which is open to new procurement models, including long-term partnering arrangements, which incentivise industry to drive down costs but allow increased profits where these are earned by improved performance;
- in addition, the Government helps sustain an attractive overall business environment, including:
  - a stable macro-economic and political environment;
  - leadership in science & technology, including by targeted MOD investment;
  - low costs;
  - Strong support industries in finance, business services, design and marketing;
  - a highly skilled and flexible labour force;
  - a transparent business environment that encourages fair competition;
  - specific support to the Defence industry, including the Defence Export Services Organisation.

xiii.   We also recognise that the bedrock of our procurement policy has to be long-term value for money. Competition is often a useful mechanism to establish this, but is not always appropriate, and needs to be used intelligently, alongside other models, considering the nature of the marketplace. The UK has increasing experience of new approaches which may apply in different circumstances, and by setting out how we approach different situations, and the various tools available, we hope in future to speed the decision-making process significantly, and pick the right tool from the toolbox first time. We also recognise the need to improve the earned profit margins available to industry based on good performance if we are to attract global investment capital into the UK defence industry.

xiv. The priority for the DIS is in ensuring that UK industry can meet the requirements of the Armed Forces, both now and in the future. Wider factors, as set out in Chapter A9, will continue to be considered in acquisition decisions. The key to ensuring that a chosen procurement strategy is most suited to the circumstances of a particular project is to expose the wider factors which impinge upon that project at the earliest opportunity, engaging relevant Government stakeholders from the outset in order to do so.

Identifying and sustaining Key Industrial Capabilities

xiv.   Every nation ideally wants to keep under its control critical defence technologies, but no country outside the US can afford to have a full cradle to grave industry in every sector, and our Armed Forces continue to benefit from the extensive range of foreign-sourced equipment currently in service. And it is readily recognised that much of the equipment procured from UK prime contractors contains non UK sourced content. We welcome the progress made in establishing understandings on security of supply and the decision to introduce an EU Code of Conduct on Defence Procurement which aims to create an effective European Defence Equipment Market. We continue to welcome overseas products, and indeed in many significant areas rely on overseas supply, with appropriate guarantees (which may include technology access to ensure we can adapt equipment to meet national requirements over time) and/or judgement that any increased risk to maintaining our operational independence is acceptable.

xv. The UK also retains a sizeable, open and broadly-based defence industry which delivers a large proportion of MOD's needs, and we welcome overseas investment, especially from companies that create value, employment,

technology or intellectual assets in the UK and thus become part of the UK defence industry. Within this strategy, we aim to tell industry very clearly where, to maintain our national security and keep the sovereign ability to use our Armed Forces in the way we choose, we need particular industrial capabilities in the UK (which does not preclude them being owned or established by foreign-owned companies). We have therefore assessed industrial capabilities against national security priorities, broken down into:

- strategic assurance (capabilities which are to be retained onshore as they provide technologies or equipment important to safeguard the state, e.g. nuclear deterrent);
- defence capability (where we require particular assurance of continued and consistent equipment performance);
- and strategic influence (in military, diplomatic or industrial terms), as well as recognising potential technology benefits attached to these which have wider value. But as the DIS makes clear, even where we wish an industrial capability to be sustained in the UK for strategic reasons, that does not necessarily preclude global competition in that sector for some projects.

PART B – Review by Industrial Sector and Cross-cutting Capabilities

B1. System Engineering

xvi.   Given that the new platforms being brought into service are likely to remain in our inventory for many years, and are increasingly complex, it is little use investing in cutting-edge science unless systems engineering capability and vital long-term knowledge is maintained. New technologies will have less benefit if the knowledge of how they might best be exploited and inserted into existing equipment has been lost. This demands a high level of systems engineering skills, at all levels of the supply chain (recognising that much of a platform's capability is delivered through its subsystems, which will often be the route to upgrading capability), sustained through the life of the equipment. The significance of this capability varies by sector, but it is generally very important for maintaining our control of how we operate our Armed Forces.

B2. Maritime

xvii.   We require versatile maritime expeditionary forces, able to project power across the globe in support of British interests and delivering effect on to land at a time and place of our choosing. To sustain this capability:

- it is a high priority for the UK to retain the suite of capabilities required to design complex ships and submarines, from concept to point of build; and the complementary skills to manage the build, integration, assurance, test, acceptance, support and upgrade of maritime platforms through-life;

- For the foreseeable future the UK will retain all of those capabilities unique to submarines and their Nuclear Steam Raising Plant, to enable their design, development, build, support, operation and decommissioning. MOD and industry must demonstrate an ability to drive down and control the costs of nuclear submarine programmes;

- We also need to retain the ability to maintain and support the Navy.

- There are a number of specific key maritime system capabilities and technologies which we should retain onshore, and the ability to develop and integrate into platforms complex maritime combat systems is also a high priority.

xviii.   In the past, we have specified that all warship hulls should be built onshore. However, the national security requirement surrounds the ability to upgrade rapidly, integrate highly complex and sensitive subsystems, and launch operations from the UK base. To sustain this requires a minimum ability to build as well as integrate complex ships in the UK, not least to develop the workforce, and to adjust first-of-class designs as they develop. At issue is the capacity required. The Future Aircraft Carrier, Type 45 Destroyer and Astute projects will keep the UK shipbuilding industry fully employed for some years (and it may not have the fabrication capacity to absorb the full programme at its peak), but from around 2016, the steady-state demand will be significantly lower. The business must be streamlined for greater efficiency and profitability. The clear trend is for fewer more capable platforms, able to incorporate upgrades as necessary to respond to new technologies and threats. The ability to do so will depend upon us working together with industry to address the fundamental issues of affordability and productivity. The industry, which is currently fragmented, needs to consolidate and refocus around a core workload which sustains key capabilities and represents a viable business. Provided our key capabilities are maintained, not all of them must be exercised onshore for every project, and the strategic need for onshore execution will be judged on a case by case basis.

xix.   We will immediately start negotiations with the key submarine companies with the aim of achieving a programme-level partnering agreement with a single industrial entity for the full life cycle of the submarine flotilla, addressing key affordability issues. The aim is to achieve this agreement in time for award of the fourth and subsequent Astute Class submarines. For Surface Ship Design & Build, within the next six months, we aim to have reached a common understanding of the core load required to sustain the high-end design, systems engineering and combat systems integration skills that we have identified as being important. We expect industry to begin restructuring itself around the emerging analysis to improve its performance, and shall build on the momentum generated by the industrial arrangements being put together on the CVF programme to drive restructuring to meet both the CVF peak and the reduced post-CVF demand. For surface ship support, we will start immediate negotiations with the industry with the aim of exploring alternative contracting arrangements and the way ahead for the next upkeep periods, which start in the autumn of 2006. Key Maritime Equipment industrial capabilities will be supported by the production of a sustainability strategy by June 2006.

B3. Armoured Fighting Vehicles (AFVs)

xx.   The AFV fleet is key to the Land Forces' military effectiveness. There are compelling advantages to retaining a UK industrial AFV capability to maintain and upgrade the capability of current and future equipment. We seek to maintain in the UK AFV Systems Engineering, Domain and Design Knowledge for though life capability management, including the ability to act as an intelligent customer for the design, development and manufacture of new AFVs and their integration into networks. We also need the intellectual ability to design, validate and interpret the results of AFV testing, though most test and evaluation facilities do not necessarily have to be on-shore. We also wish the UK defence industry to be able to design, build and integrate onto the platform AFVs' critical subsystems, including electronic architecture, sensors and integrated survivability solutions. We also need to be able to repair and overhaul AFVs onshore, and we need the industry to be able to respond quickly, including through deployed support on operations. For future projects, we need industry to deliver the complex system of systems that will make up the Future Rapid Effects System (FRES) fleet.

xxi.   It is questionable whether any single company has the ability or expertise to provide all elements of the FRES capability cost-effectively. The most likely solution will be a team, led by a systems integrator with the highest levels of systems engineering, skills, resources and capabilities based in the UK, in which national and international companies cooperate to deliver the FRES platforms, including the required subsystems.

xxii.   The UK AFV industry has consolidated so that BAE Systems Land Systems (LS) is the supplier of 95% of our current inventory. We need to manage this in-service fleet through life whilst still retaining access to best of market products at subsystem level. Building on discussions already set in train, we will work hard with the company to give effect to the long-term partnering arrangement required to improve the reliability, availability and effectiveness through-life of our existing AFV fleets. We intend to establish a joint team early in 2006 to establish a business transformation plan underpinned by a robust milestone and performance regime. We expect to see a significant evolution of BAE Systems Land Systems both to deliver AFV availability and upgrades through life, and to bring advanced land systems' technologies, skills and processes into the UK. If successful in their evolution, BAE Systems will be well placed for the forthcoming FRES programme.

B4. Fixed wing

xxiii.   Air power continues to offer the ability to transform the battlespace, utilising its inherent attributes of reach and speed to enable strategic operational and tactical agility. We are introducing two new, highly sophisticated manned combat fast jets, Typhoon and the Joint Combat Aircraft, which are intended to last for more than 30 years. Current plans do not envisage the UK needing to design and build a future generation of manned fast jet aircraft beyond these types. However, precisely because the current fleet and the new types we are introducing are likely to have such long operational lives, we need to retain the ability to maintain and upgrade these types for a considerable period.

xxiii.   The focus must shift to through-life support and upgrade and what is required to sustain this critical capability in the absence of large-scale manufacture. MOD has been working closely with BAE Systems, as the UK's only supplier of fast jets, for some time to understand these mutual challenges, which are likely to impact on the UK industrial footprint, in particular around BAE Air Systems' four main production sites. We intend to continue to work together to explore how a long term partnering arrangement for the through-life availability of a significant proportion of the fixed-wing fleet might be delivered to sustain these capabilities and deliver improved value for money. We aim on working during 2006 to develop the solution – which will be challenging given the scale of the transformation that is required – and to implement it from 2007.

xxiv.   We and industry share a close alignment of interest in UAV and UCAV technology. Although at present we have no funded UCAV programme, targeted investment in UCAV technology demonstrator programmes would help sustain the very aerospace engineering and design capabilities we will need to operate and support our future aircraft fleet. Such investment would also ensure that we can make better informed decisions which will need to be taken around 2010-2015 on the future mix of manned and unmanned aircraft. Additionally, UK industry will have the opportunity to develop a competitive edge in a potentially lucrative military and civil market. We intend to move forward with a substantial joint Technology Demonstrator Programme in this area. We hope that appropriate arrangements will be in place to allow this to proceed in 2006.

xxv.   Our plans to retain onshore the industrial capabilities required to ensure effective through-life support to the existing and planned fast jet fleet – and to invest in developing UCAV technology – will also provide us with the core industrial skills required to contribute to any future international manned fast jet programme, should the requirement for one emerge. This recognises both the uncertainty of our very long term requirements – with the possibility that we shall want to replace elements of the Typhoon and Joint Strike Fight fleets with manned aircraft – and that we should avoid continuing to fund industrial capabilities for which we have no identified requirement.

xxvi. Critical mission systems, including electro-optical (EO) sensors, radar, Electronic Support Measures (ESM) and Defensive Aids Systems (DAS) are also significant areas where we wish to retain onshore capability and where suppliers must be able to work with the prime contractor and be rewarded for developing new solutions.

xxvii. Our need to retain a minimum level of onshore capability does not necessarily mean that we will need to support all aspects of our aircraft in the UK. For Typhoon, we will work with our partners to create a better and more efficient business model for the aircraft's support and upgrades, ensuring that we retain onshore our ability to satisfy our sovereign requirements over its lifetime. Clearly, BAE Systems, and, for the engines and mission systems respectively, Rolls-Royce, Smiths Aerospace and Selex Sensors and Airborne Systems will have a significant role to play in this..

xxviii. For the Joint Strike Fighter, the through-life support of the UK aircraft will be provided from the Lockheed Martin Global Support System which is being established on a co-operative basis amongst the nine JSF partner nations. As part of this performance based arrangement, the UK also intends to establish sovereign support capabilities which would provide, in country facilities to maintain, repair and upgrade the UK fleet and an Integrated Pilot and Maintainer Training Centre. Our aim is that BAE Systems as a key JSF Industry partner to Lockheed Martin will provide these support services in the UK under a Team JSF badge. There is no fundamental defence requirement for a JSF Final Assembly and Check Out (FACO) facility, although an ongoing joint study between MOD, DTI and BAE Systems, due to conclude in early 2006, is seeking to assess whether a UK FACO is necessary to preserve essential engineering skills within BAE Systems and would be a cost effective and affordable solution.

xxix. There is no sovereign requirement to sustain an indigenous capability in large and training aircraft. We will continue to need, however, the systems engineering and design skills and Intellectual Property Rights for the integration of new mission systems, avionics and defensive aids into these platforms.

B5. Helicopters

xxx. Helicopters are inherently responsive, adaptable and flexible, and contribute to a variety of military tasks. They can operate in a very wide range of combat and environmental conditions, and will often be an essential part of a balanced expeditionary force.

xxxi. The helicopter sector has similar characteristics to the AFV sector – a high concentration of knowledge relating to the existing fleet, but a healthy international competitive environment. AgustaWestland's systems engineering capability needs sustainment to maintain our ability to support and upgrade the current fleet.

xxxii. Our preferred solution is to invest in the Future Lynx product, currently undergoing detailed capability and value for money assessment, to meet our Battlefield Reconnaissance and Surface Combatant Maritime Helicopter requirements and sustain the necessary Design Authority capability at the company in the short to medium-term. We intend to promote a more open, predictable but demanding partnered relationship with the company, to provide better value for money and reduce their reliance on our investment to sustain the design engineering skill-base, and accordingly intend to finalise a Strategic Partnering Agreement with AgustaWestland by Spring 2006. We will continue to look to the vibrant and competitive global marketplace to satisfy our future helicopter requirements (including for support). We also wish to keep different levels of capability onshore in rotorblades, mission systems, survivability, vibration management and electronic architecture.

B6. General munitions

xxxiii. Recent operations have clearly demonstrated that despite the increases in technology, modern warfare, particularly on the ground, requires highly trained and motivated service personnel to engage in combat at a very personal level. It is in such engagements that quality general munitions are essential to provide the volumes of fire and the 24 hour, all weather capability required to suppress, neutralise and demoralise enemy forces. It is essential that we retain onshore the Design Authority (DA) role and its underpinning capability for munitions manufactured. We also require the ability to develop munitions for specific purposes to match our doctrine, and maintain an intelligent customer capability for non-UK designed munitions. A robust through-life management capability onshore is vital. It is also essential that we retain a proof and surveillance capability onshore for UK designed munitions as well as at least a minimum munitions disposals capability. We should also retain onshore the UK's insensitive munitions and related energetic materials capability, which are world-class. But we do not consider it necessary to retain all aspects of bulk explosives manufacture in UK and would be prepared to source small arms ammunition offshore if security of supply could be guaranteed; it is presently questionable given potential undercapacity in global supply.

xxxiv. In this sector, BAE Systems has the majority of the existing business, but there remain niche capabilities abroad and elsewhere in the UK which may meet future needs. We have therefore adopted a partnership with BAE Systems and are considering ways in which we can rationalise the through-life management of munitions, without ruling out the prospect of global competition for future projects at this stage. We also have partnering agreements with other suppliers (Rheinmettall and Wallop Defence Systems) in niche areas. We will reach further conclusions on how best to sustain our required access to general munitions in summer 2006.

B7. Complex weapons

xxxv. Complex Weapons provide our Armed Forces with battle winning precision effects. The UK is making a significant investment in the upgrade and development of complex weapons, which peaks at just over £1BN next year and will reduce by some 40% over the next five years following the delivery of Storm Shadow and Brimstone. There is, apart from the Meteor programme, little significant planned design and development work beyond the next two years. This will present a substantial challenge to the industry.

xxxvi. There are some types of complex weapon that we have bought from overseas in the past, and we would be prepared to source future torpedoes from abroad provided we retain the capability to support the current inventory, write tactical software, and design and integrate homing heads. However, we would wish to maintain the ability to design, develop, assemble, support and upgrade other complex weapons, which is a complex task requiring a number of critical and sensitive underpinning capabilities. We also see the potential of Directed Energy Weapons.

xxxvii. The fragility of the wider UK industrial base is such that open international competition could put the sustainment of key industrial capabilities at risk. We intend to work with all elements of the onshore industry over the next six to twelve months to establish whether – and if so how – we can achieve a sustainable industry that meets our requirements in a value for money fashion. There is potential for industrial rationalisation and consolidation and we will need to work with other European governments to identify whether a coordinated approach to sustain a viable industrial base is possible. But this will not be to the exclusion of US-owned companies, in particular those who have established a firm foothold in the UK.

B8. Command, Control, Communication and Computers, Intelligence, Surveillance, Target Acquisition and Reconnaissance (C4ISTAR)

xxxviii.   This is a very significant area where we assume sustained expenditure. It will be the C4ISTAR related capabilities that will help underpin the overarching Network Enabled Capability essential to the continued transformation of our capability, by providing the technology to deliver agile, networked and informed Armed Forces.

xxxix.   Much of the innovation is driven by the civil sector and we are in general a relatively minor customer in a market where the pace of technological change creates its own set of unique pressures. To maintain national security, we need to maintain in the UK specific industrial capabilities, including:

- High grade cryptography and associated information assurance capabilities;

- A continued ability to understand, integrate, assure and modify mission critical systems.

as well as intelligent customer status and a research and development base supported by a manufacturing capability in specific areas.

xxxx.   There are a number of healthy companies with the requisite skills in the UK, and given civil opportunities in this sector and a large number of planned projects, competition by project seems sustainable for the foreseeable future. However, maintaining a cryptographic capability currently requires a specific strategy to sustain an end-to-end design, development and manufacturing capability. We are working with other government departments to generate better coherence across Government, and increase industry's visibility of the total opportunities.

B9. Chemical, Biological, Radiological, and Nuclear Force Protection

xxxxi.   We are committed to maintaining the UK's political and military freedom of action despite the presence, threat or use of CBRN weapons, and this is an area in which significant increases in investment are currently planned. We need the UK industrial base, which is a world leader in this field, to deliver intelligent supplier capabilities, systems engineering, specific technology research, as well as the supply of certain raw materials and the manufacture of medical countermeasures.

xxxxii.   CBRN protection requirements have for some time been met through a healthy competitive industrial market place. We will explore however the potential costs and benefits of partnering, however, particularly with the four main industrial players in the UK (Smiths Detection, General Dynamics UK, Serco Assurance and EDS), to see whether other acquisition models could allow us to achieve rapid and innovative acquisition and achieve better value for money.

B.10 Counter terrorism (CT)

xxxxiii.   Given the nature of the international terrorist threat, capabilities previously needed in specialist areas and in Northern Ireland are increasingly becoming required across the Armed Forces. This reinforces the importance of the counter-terrorism sector, and provides greater opportunities for both industry and MOD to become more cost-effective in the CT field.

xxxxiv.   Although there are aspects of the technology base within the development, manufacture and sustainment of a CT system that need to be retained within UK industry, it is primarily within the areas of systems engineering (including design and development), testing and evaluation, and system packaging that the MOD needs to be able to maintain critical elements of its CT capability onshore. We believe there is no urgent remedial action required to sustain these industrial capabilities.

B.11 Technology priorities to enable defence capability

xxxxv.   To support the industrial capabilities identified across the sectoral analysis there are a number of areas in which the UK must sustain existing technological strengths or where we should, resources permitting, consider developing our expertise. There are other technologies showing promise across a range of defence applications that may have either a large impact on specific defence capabilities or a more widespread impact across many aspects of defence. These are provisionally identified in the DIS, but we recognise we will need further work in 2006 to inform our research and technology priorities.

B.12 Test & evaluation (T&E)

xxxxvi.   T&E is vital to the development, introduction into service and through-life support of the equipment used by our Armed Forces. It contributes to a variety of activities which reduce risk to our Armed Forces. We use a mixture of in-house, Government Owned Contractor Operated (GoCo) and commercial T&E facilities in the UK to support the acquisition and sustainment of military capability. The majority of MOD T&E sites operated on our behalf by QinetiQ under the Long Term Partnering Agreement (LTPA). All these capabilities are kept under constant review to ensure that they continue to meet our T&E requirements and to identify potential rationalisation or efficiency opportunities.

xxxxvii.   In some cases a UK based T&E capability is essential for, amongst other things, certain quality assurance, safety or operational security needs and sovereignty of access. In other cases the important element is to retain the ability to direct, understand, analyse and verify T&E results rather than actually conduct testing on-shore, subject to certain safeguards including security of supply. We will work with industry to identify where such distinctions can be safely made. Our current strategic intent in the medium term is to retain T&E capability within the UK, but to look for overseas cooperation where appropriate. Work in the European Defence Agency may lead, in due course, to a longer-term strategy to consolidate T&E capabilities across Europe.

PART C: Implementing the Defence Industrial Strategy

xxxxviii.   The DIS also presents real and fundamental challenges to the Ministry of Defence. The strategy will not deliver unless the whole of the defence acquisition community, including industry, are able to make the necessary shifts in behaviours, organisations and business processes.

il.   The basic principles of Smart Acquisition still hold true and are a strong foundation from which to take forward the DIS. But our future approach to acquisition must be built around achieving primacy of through life considerations; coherence of defence spend across research and development, procurement and support; and successful management of acquisition at the departmental level. Our detailed implementation plan has specific initiatives to address the objectives of achieving:

- primacy of through-life considerations;
- coherence of defence spread accross research, development, procurement and support;
- sucessful management of acquisition at the Departmental level.

l.   The measures identified under these headings are necessary
to improve our acquisition performance. But they may not be sufficient.
We will appoint a senior official to review our current acquisition
construct and recommend changes across the MOD's business with
final recommendations by May 2006 for early implementation.

li.   We will be looking for parallel commitment
from industry in the following areas:

- planning more effectively and jointly for the long term,
  embracing the vision of through-life capability management
  to meet our requirements cost-effectively;

- investing in growing and maintaining a high-quality
  systems engineering capability within the UK;

- promoting greater interaction and collaboration between
  MOD, prime contractors, SMEs and the universities to stimulate
  innovation in science, technology and engineering;

- encouraging trust, openness, transparency and
  communication with MOD at all levels;

- embracing open systems architecture principles and incremental
  acquisition approaches throughout the supply chain;

- working jointly to foster better understanding of each others'
  objectives and business processes, including a greater commitment
  to joint education, staff development and interchange opportunities.

lii.   We will keep the progress of this work, and the extent to which
real change is being demonstrated on the ground, under review within
the MOD, through the Acquisition Policy Board reporting to the Minister
for Defence Procurement. We will want formally to review progress with
the National Defence Industries Council regularly. We will also review this
Strategy as a whole once every Comprehensive Spending Review period.

section A

Strategic Overview

A

Strategic Overview

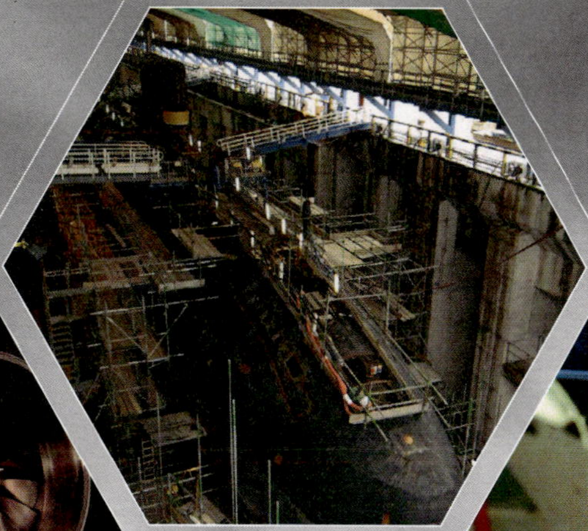

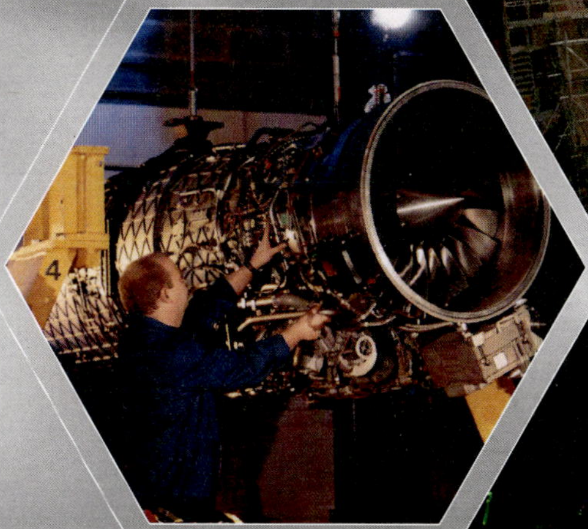

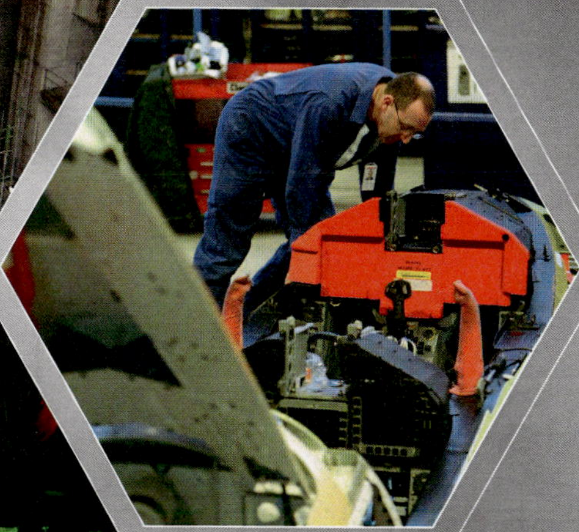

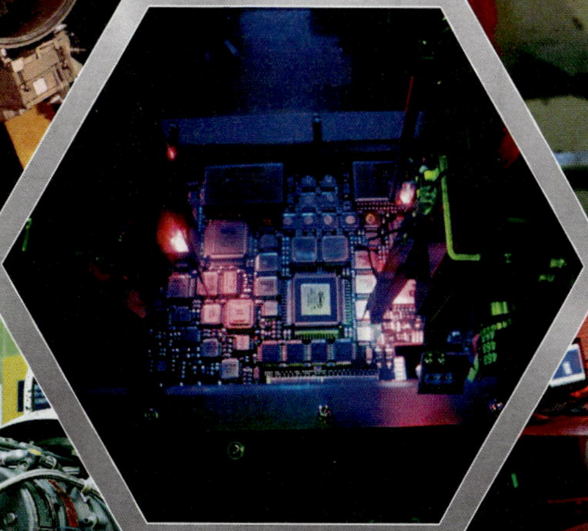

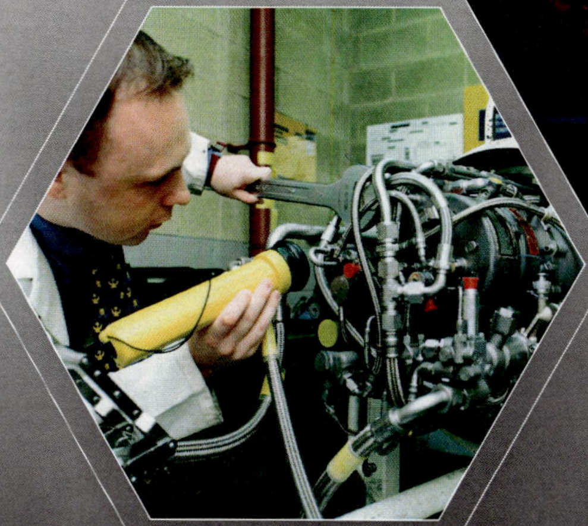

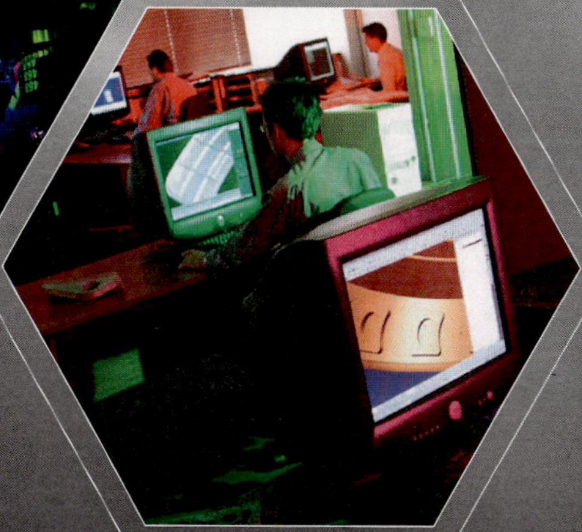

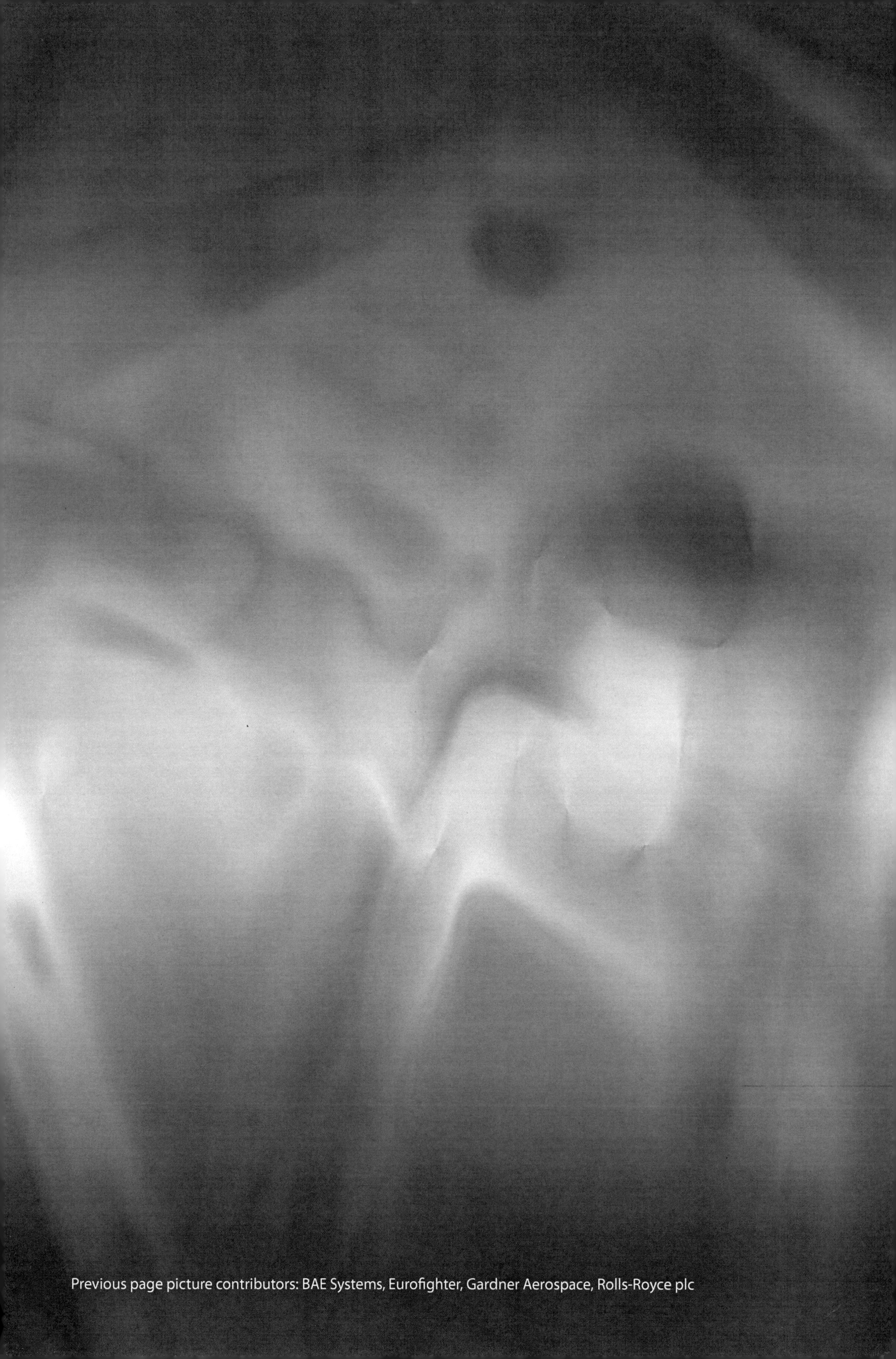

Previous page picture contributors: BAE Systems, Eurofighter, Gardner Aerospace, Rolls-Royce plc

Introduction

A1.1   The global security environment and the policy context in which the Armed Forces operate has undergone substantial changes over the past fifteen years. No longer do we face an imminent and existential threat from a hostile superpower; but in its place have developed new and more complex challenges. The 1998 'Strategic Defence Review' (SDR) recognised the significance of this strategic shift and established a clear policy requirement for flexible, expeditionary Armed Forces able to undertake a wide range of tasks at distance from the UK. Since then we have continued to review our implementation of this broad policy direction, responding to rapid evolution of the strategic environment. This is evident in the series of further policy papers including the 2002 'SDR New Chapter', the 2003 'Delivering Security in a Changing World' White Paper and the 2004 'Future Capabilities' Command Paper. These have driven continuous change in the roles, doctrine, size and shape of our Armed Forces.

A sniper with Support Company, 1st Battalion, The Irish Guards, covers Royal Engineers extinguishing an oil well fire, near Basra.

A1.2   In parallel, the defence industry has evolved. The defence sector has not been immune to the forces of globalisation and inter-dependence that have characterised the wider world economy over the past thirty years. Far from it: defence companies are now transnational, with the need to attract and retain investors - forcing increased efficiency, restructuring and rationalisation and an increased focus on which markets secure best returns to shareholders.

A1.3   Nowhere is this more true than in the UK, which has for the past fifteen years operated one of the most open defence markets in the world. Exposing UK defence suppliers to the rigours of competitive pressure has reaped huge dividends: for the Armed Forces, in ensuring that they have access to the best military hardware on offer; to the tax payer, in reducing substantially the costs of meeting these requirements; and to UK industry itself, which having been incentivised to increase productivity in the domestic market has been so successful in winning business abroad.

A1.4   We have now reached a crossroads. We are seeing a shift away from platform orientated programmes towards a capability-based approach, with corresponding implications for the demand required of the traditional defence industrial base. Although we are in the middle of a substantial transformation, involving a series of major new platforms (including the future Aircraft Carriers, Type 45 Destroyers, new medium-weight armoured fighting vehicles, and the A400M, Typhoon and Joint Strike Fighter) we expect these platforms to have very long service lives. This means the future business for the defence industry in many sectors will be in supporting and upgrading these platforms, rapidly inserting new technology to meet emerging threats, fulfil new requirements and respond to innovative opportunities, not immediately moving to design the next generation. In parallel, the extent of industrial rationalisation means that sustaining competition to meet domestic requirements is increasingly difficult. In several sectors, following the entry into service of major projects, there will be substantial overcapacity in production facilities in the UK defence industry in a few years' time.

A1.5   And as we look to non-UK sources of supply, whether at the prime or sub-systems level, we need to continue to recognise the extent to which this constrains the choices we can make about how we use our Armed Forces - in other words, how we maintain our sovereignty and national security. We need to be clear that the international nature of the defence business is such that companies now have more choice than ever before about the markets in which they choose to operate, which secure the best returns for shareholders, and where to base their operations. And we need to be clear about how the nature of global capital markets shapes these commercial judgements. If we do not make clear which industrial capabilities we need to have onshore (and this includes those maintained by foreign-owned defence companies), industry may make independent decisions which would lead to the disappearance of indigenous capabilities required to maintain our national security.

A1.6   Equally, we do not seek to restrict the scope for international cooperation and competition where this is appropriate, and we cannot afford to maintain a complete cradle-to-grave industrial base in all areas. As industry has told us, greater clarity is therefore needed urgently on which capabilities must be retained onshore, and which by implication can be met from a wider market.

A1.7   These factors necessitate a fresh consideration of how best we should seek to engage with the marketplace in order to meet our requirements; and drive the need for a Defence Industrial Strategy (DIS).

The aim of the Defence Industrial Strategy

A1.8   The DIS flows from the Defence Industrial Policy (DIP) published in 2002. Like that policy, it is 'driven by the need to provide the Armed Forces with the equipment which they require, on time, and at best value for money for the taxpayer.' The DIS is thus one of many contributions to the wider aim of ensuring that the capability requirements of the Armed Forces can be met, now and in the future.

A1.9   The DIS is thus one of many contributions to the wider aim of ensuring that the capability requirements of the Armed Forces can be met, now and in the future.

A1.10   The contribution that a DIS can offer towards that overriding aim is in promoting a sustainable industrial base, that retains in the UK those industrial capabilities (including infrastructure, skills, tacit knowledge, Intellectual Property (IP) and capacity) needed to ensure appropriate sovereignty and/or contribute to co-operation with allies, to ensure our national security, but allows us to benefit from products on the broader international market where appropriate, to maximise our Armed Forces' cost-effectiveness. Our interaction with this industrial base must provide good value to the taxpayer and good returns to shareholders based on delivery of good performance, and be consistent with broader security and economic policy. This explicitly includes the need to continue to make the UK a great place to establish, grow and invest in high-technology defence businesses.

A1.11   This recognises that while the industrial base from which we procure is global, sourcing from the UK in certain areas will be essential for our long-term interests. The UK industrial base needs to be sustainable, which means taking a long-term perspective, considering all the levers available to government, and taking into account the impact of different procurement choices on specific sectors or industry as a whole. We recognise the need to generate fair returns in business. Consistency with broader security and economic policy in the contribution includes issues like:

● the importance of preventing WMD related, and undesirable conventional, proliferation;

● consistency with the Government's spending plans, and a recognition that priorities may change in the future;

● consistency with other government initiatives and strategies, such as the Government's Manufacturing Strategy and our Technology Strategy, which also affect parts of the UK defence industry[1].

A1.12   This acknowledges the over-capacity that exists - when viewed against our likely future requirements - in several sectors of the industrial base; and recognises the likelihood that some further rationalisation is inevitable (and in some cases is already being actively considered by certain companies). As such, it provides a mechanism for shaping and managing that process to the benefit of both the customer and its suppliers.

Objectives

A1.13   To deliver this aim, the DIS:

● gives a strategic view of defence capability requirements going forward (including new projects, but also the support and upgrade of equipment already in-service), by sector[2]. Part of the strategic view is specifying, in order to meet these, which industrial capabilities we would wish to see retained in the UK for defence reasons. We aim to communicate the overall view to industry as clearly as possible, recognising that plans change as the strategic or financial environment changes;

● gives further detail on the principles and processes that underpin procurement and industrial decisions;

● where there is a mismatch between the level of activity our own plans (and export/civil opportunities) would support and that required to sustain desired capabilities, investigates how we might with industry address that gap, within the bounds of affordability.

A1.14   As a result, industry will be better able to make informed investment decisions, and industry and Government can focus on improving delivery and productivity. Industry should note that if an industrial capability is not specified as a strategic priority for retention in the UK, this does not necessarily mean UK companies cannot win business in those areas; we may still have important projects there.

The scope of the DIS

A1.15   We began work last year to construct a framework or matrix of key technologies which, for reasons of national security, broader defence interest or wider economic benefits, we might wish to retain in the UK, so that these could be explicitly taken into account in future procurement decisions. But we now intend to move beyond saying what we care about, to how this may be fostered and maintained, not least because it is clear that several branches of industry are considering making substantial and time-sensitive independent restructuring decisions in the near future.

A1.16   Given these drivers for early clarity, we needed to act quickly. There are three levels to this strategy:

● promoting an overall business environment which is attractive to defence companies and investors;

● identifying key industrial capabilities which are important to Defence to retain in the UK industrial base to maintain appropriate sovereignty (see further below), with sustainment strategies where these seem at risk;

● explaining how, in decisions on individual projects, we take into account industrial and other factors.

---

[2] Although we are increasingly thinking and structuring ourselves in capability or technology terms we recognise that industry is structured by sector or product, and thus we have to articulate our needs in this way within this document.
[3] i.e., focus on military equipment and operational services (such as the Strategic Sealift Service and the Future Strategic Tanker Aircraft) rather than indirect support services (such as regional prime contracting, non-operational Information Technology, facilities management) which also fall within our definition of the UK defence industry ('Economic activity that is supported by UK MOD spending and exports of defence goods and services').

---

[1] The DIP states that the UK defence industry 'embraces all defence suppliers that create value, employment, technology or intellectual assets in the UK. This includes both UK and foreign-owned companies'. 'The UK defence industry should therefore be defined in terms of where the technology is created, where the skills and the intellectual property reside, where jobs are created and sustained, and where the investment is made.'

A1.17   For the second of these, developing a strategic view has meant focusing on key sectors which have direct impact on key defence outputs[3] and where a) we anticipate potential restructuring in any case; and/or b) which are strategic priorities for our future capabilities. In each area, we have only drilled-down to the extent necessary. In Part B, therefore, we have covered the following sectors, in all cases for support and upgrade as well as initial acquisition:

- Submarines and surface ships
- Armoured Fighting Vehicles
- Fixed wing aircraft, including UAVs
- Helicopters
- General munitions
- Complex weapons
- C4ISTAR
- CBRN Force Protection
- Counter Terrorism

A1.18   In addition, there are three separate chapters on the very important cross-cutting issues of **Technology Priorities, Systems Engineering** and **Test & Evaluation** (T&E).

Full deck operations during handover with HMS INVINCIBLE.

A1.19   Taking forward this strategy is likely to mean internal change within Government and MOD too; this is covered with Part C, on change and implementation. But we precede these sections with a **Strategic Context** in Part A, which includes the military context which drives our equipment needs; analysis of the changing shape of the global and UK defence marketplace; an overview of the features which make the UK attractive to Defence companies, researchers and investors; and explains how we take industrial and other wider factors like export potential into account, including in deciding between competitive and other procurement strategies.

Guiding principles

A1.20   Against this background, and in order to set the framework for the DIS, six guiding principles have been developed.

Appropriate sovereignty

A1.21   We must maintain the appropriate degree of sovereignty over industrial skills, capacities, capabilities and technology to ensure <u>operational independence</u> against the range of operations that we wish to be able to conduct. This is not 'procurement independence', or total reliance on national supply of all elements, and will differ across technologies and projects. It covers not only being assured of delivery of ongoing contracts, but also the ability to respond to Urgent Operational Requirements (UORs) (taking into account other customers' likely demands at the same time), where systems engineering skills amongst others may be important, and to support in-service equipment. In many, even high priority areas, we can, and do, rely on overseas sources, and have made progress in recent years in developing

increased assurances of security of supply, but there are critical areas where not maintaining assured access to onshore industrial capabilities would compromise this operational independence and hence our national security. The extent to which we feel comfortable sourcing defence equipment from overseas is also a function, amongst other things, of our ability to negotiate with other nations arrangements to share the technologies required to support such capabilities through-life and adjust them to our national requirements as necessary. Such national security considerations are also relevant where we need to retain sovereignty due to the extreme security sensitivity of the technology concerned or for legal reasons; where specific UK capabilities give us important strategic influence, in military, diplomatic or industrial terms; and in some cases, where retention is necessary to maintain realistic global competition - in other words, where we are not prepared to risk dependency on an overseas monopoly which could in time frustrate our ability to maintain our freedom of military action.

A1.22   At the same time, we must be prepared to exploit the opportunities engendered by co-operative arrangements with others, where it makes operational and economic sense to do so; recognise that increasing mutual dependence on supply from other states is in many cases not only happening but desirable; and continue to encourage inward investment from overseas. We articulate in this strategy some of the policy tools available to maintain the 'national' status of different technologies, recognising that the legal and policy frameworks for defence markets in Europe and the United States are evolving. We of course also recognise the broader attractions of the UK for business, including macro economic stability and the support available to UK industry and science and technology from other government departments and agencies.

Through-life capability management

A1.23   There is a general shift in defence acquisition away from the traditional pattern of designing and manufacturing successive generations of platforms - leaps of capability with major new procurements or very significant upgrade packages - towards a new paradigm centred on support, sustainability and the incremental enhancement of existing capabilities from technology insertions. The emphasis will increasingly be on through-life capability management, developing open architectures that facilitate this and maintaining - and possibly enhancing - the systems engineering competencies that underpin it. The attractions for industry should, in general, include longer, more assured revenue streams based on long-term support and ongoing development rather than a series of big 'must win' procurements.

Maintaining key and rapid industrial capabilities and skills

A1.24   In those areas where reduced UK and export market opportunities cannot any longer provide a sustainable production profile, the DIS needs to address the challenge of maintaining key industrial capabilities. Often, this may be about sustaining and developing small pools of expertise or knowledge (including tacit knowledge) within or across the supply chain rather than focused on particular facilities or existing technologies. We need to identify these as far as possible, and develop appropriate options to maintain them.

Intelligent customers-intelligent suppliers: the importance of systems engineering

A1.25   Notwithstanding that some key knowledge potentially rests in very small teams, in many (though not all) sectors, the ability to understand and sometimes manage the complexities, challenges and costs associated with overall management of design, manufacture and upgrade remains a general requirement. This is important at a number of different points in the acquisition cycle: we need to preserve the capacity for sensible industry-MOD conversations when a capability is in concept phase and a number of potential technologies to deliver it are being considered; systems engineering at various levels is critical for the successful acquisition of complex projects and programmes; and in-life upgrades, including UORs, often require deep systems engineering skills and knowledge, often relating to systems brought into service many years before. But the depth of design for producability and systems engineering capability, and the concentration of effort needed in the UK supply base will vary throughout the CADMID[4] cycle and from sector to sector, and needs to be considered accordingly.

Value for Defence

A1.26   Driving long-term best value for money lies at the heart of our Defence acquisition policy. But despite the opportunities to exploit the internationalization of the defence supply chain to generate cost savings and other advantages, we recognise the benefits that flow from the existence of a healthy, competitive and dynamic national industry (whether in terms of the amortisation of overheads associated with export sales or the risks of being subject to monopoly power should we have to look exclusively overseas for some requirements). As the DIP made clear, where defence procurement decisions impact on wider Government, this will be considered and the DIS will include understanding of the role of Government and industry, in setting a framework which maximises value and contributes to the national Science and Technology base.

Change on both sides

A1.27   Industry is likely to change as a result of taking the DIS' conclusions into account. We, along with the wider Government, will need to change too. As well as considering structural and cultural changes, we have set out more clearly our future plans. We have also explained how we can identify situations where, unusually, competition at a particular point or level is not the best solution to drive innovation, encourage investment and produce a fair price; and indicate how, in those situations, value for money is scrutinized, incentivised and protected. We have also set out the improvement in performance we expect from the supply side.

National Defence Industries Council (NDIC).

[4] *Concept, Assessment, Demonstration, Manufacture, In-Service, Disposal*

A2

Delivering security in a changing world

A2.1   The December 2003 Defence White Paper 'Delivering security in a changing world' set out the Government's analysis of the future security environment, the implications for defence, our strategic priorities and how we intend to adapt our planning and force structures to meet the most likely threats and challenges. The White Paper explained the need to adapt to the more pronounced threats presented by international terrorism, the proliferation of Weapons of Mass Destruction (WMD) and the challenges posed to the international community by weak and failing states. It also explained how Defence should exploit the opportunities presented by effects-based planning and operations together with highly networked and adaptable forces across all three Services.

Chinook helicopters delivering aid after the devastating earthquake in Pakistan.

A2.2   The White Paper formed the policy baseline for the 'Future Capabilities' Command Paper of July 2004, which set out the force structure and capability changes required to respond to these changes in the security environment. 'Future Capabilities' indicated our intention to continue the process of modernising Defence, investing our resources in the capabilities and structures which provide flexible and adaptable high quality Armed Forces, properly equipped to deal with the challenges and threats of the future.

A2.3   The documents set out a revised set of assumptions and implications to underpin future Defence planning. These include:

- the need to defend the UK, protect our interests overseas, counter the threats from the proliferation of WMD and international terrorism, and deal with the consequences of weak and failing states requires a clear focus on projecting force, further afield and

even more quickly than has previously been the case. This places a premium on the deployability and sustainability of our forces;

- that we should plan to be able to operate in six core regions: the Near East, the Gulf, sub-Saharan Africa, and South Asia, and in and around Europe. However, counter-terrorism and counter proliferation operations in particular will require rapidly deployable forces able to respond swiftly and achieve precise effects in a range of environments across the world;

- that the force structure should be rebalanced and optimised to meet the demands of up to three concurrent small and medium scale operations;

- that we should retain the flexibility to reconfigure for less frequent, but more complex and demanding, large scale operations, while concurrently conducting a small-scale peace support operation. For large scale operations we will not need to generate the full spectrum of military capabilities as the most demanding operations could only conceivably be undertaken alongside the US, either as a NATO operation or US-led coalition. Where the UK chooses to be engaged, we will wish to be able to influence political and military decision-making throughout the crisis, including during the post-conflict period. To exploit this effectively, our Armed Forces need to be interoperable with US command and control structures, match the US operational tempo and provide those capabilities that deliver the greatest impact when operating alongside the US;

- that we should maintain a broad spectrum of maritime, land air, logistics, C4ISTAR[1] and Special Forces capability elements to ensure we are able to conduct limited national operations, and be capable to lead and act as the framework nation for coalition operations where the US is not involved;

- that we must continue the transformation of our forces to concentrate on speed, precision, agility, deployability, reach and sustainability. Key to this is our ability to exploit the benefits of Network Enabled Capability, precision munitions and the development of effects-based planning and operations;

- that we need capabilities which can rapidly come together to achieve specific military effect and then rapidly adapt with other capabilities to achieve what is required by the next operation. By doing so, decisive military effect may be achieved through a smaller number of more capable, linked assets acting quickly and precisely to achieve a desired outcome.

A2.4   Our future forces must be strategically agile in order to respond to changing needs and circumstances. Agility comprises four key attributes:

- **Responsiveness** - Those capabilities that underpin our speed of response: Intelligence, Surveillance and Reconnaissance (ISR), force projection, and logistic support. Early application of military capability is likely to deliver strategic effects more quickly and economically. Therefore, improved strategic reach, precision and endurance are desirable characteristics.

---

[1] *Command, Control, Communications, and Computers, Intelligence, Surveillance, Target Acquisition and Reconnaissance.*

● **Robustness** - Future equipment must operate with increased reliability and availability. Force elements must be able to maintain sustained, high tempo operations against a range of diverse but effective opponents. We will require combat and support capabilities that are versatile across the full spectrum of operations.

● **Flexibility** - People, structures and equipment should be capable of conducting a wide range of Military Tasks with the flexibility to reorganise and re-role. Wherever possible, niche capabilities that have limited utility across the range of Military Tasks should be avoided.

● **Adaptability** - Personnel and units should be well-trained and have high utility equipment so they can adapt to changing circumstances. An important aspect of adaptability is being able to readily introduce modifications arising from new technology.

A2.5   To achieve our mission within a challenging strategic environment will require flexibility from across Defence, from our equipment, force structures and people. We must adapt to stay ahead of potential adversaries and be prepared to make tough decisions to ensure that our forces and equipment deliver the required capabilities. Force structures will need constantly to evolve as we seek to exploit new technologies, equipment and techniques to improve capability and respond to the changing strategic environment.

A2.6   Within a resource-constrained environment we must aim to maintain a technological and capability edge over our likely adversaries but not pursue the best for the best's sake. We also require incremental acquisition processes with procedures that align equipment requirements and specifications strictly to the current and expected future threat, recognising and accepting the value and potential cost of flexibility and incremental upgrades. We need to develop standard and Urgent Operational Requirement (UOR) procurement processes that are flexible, agile and responsive and minimise the lead time between the emergence of new technology or threat capabilities, and the delivery of new equipment or enhancements to existing capabilities. 'Just in Time' investment to fine tune equipment and capabilities rapidly with the latest technology in readiness for specific operations will be a necessary and increasingly important feature of operations. Adaptive and modular architectures could support system resilience and allow for systems evolution and insertion of new hardware and software. In the context of the latter, commercial-off-the-shelf (COTS) technology has much to offer but potential risks in reliability, supportability and security must be adequately addressed.

A2.7   At the heart of the force structure and capabilities modernising programme is Network Enabled Capability (NEC). NEC is about the coherent integration of sensors, decision-makers and weapon systems along with support capabilities. NEC will enable the UK to operate more effectively in the future strategic environment through the more efficient sharing and exploitation of information within the UK Armed Forces and with other coalition partners. This will lead to better situational awareness across the board, facilitating improved decision making, and bringing to bear the right military capabilities at the right time to achieve the desired military effect. This enhanced capability is about more then equipment; it includes exploiting the benefits to be obtained from transformed doctrine and training, and optimised command and control structures. The ability to respond more quickly and precisely will act as a force multiplier enabling our forces to achieve the desired effect through smaller numbers or more capable linked assets. In summary, the emphasis is no longer on quantity as a measure of capability.

During Operation TELIC the RAF employed only about 70% of the number of fast jets used in Operation Granby (the first Gulf conflict), but to much greater overall effect

A Land Rover and trailer await loading onto a RAF C-130 Hercules at Skopje military airport at the end of Operation Essential Harvest (a 5000 strong multinational operation that collected over 3000 weapons in the Balkans region).

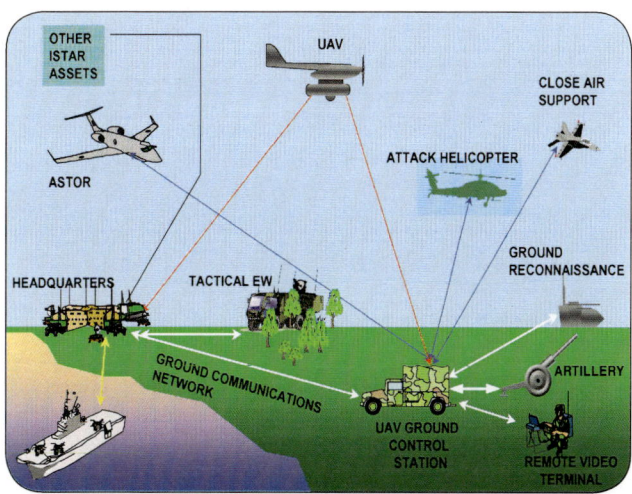

An example of Network Enabled Capability (NEC).

A2.8   Maritime - The sea provides a degree of security, without reliance on the territory of others, for projecting power across the battle space to strengthen our military advantage. Furthermore, protection of our maritime supply routes is essential in securing our ability to trade globally thus critical to the UK's wealth and security. To meet the challenges of the future, the Royal Navy (RN) will need to be an increasingly versatile, network-enabled expeditionary force, continuing to operate on, under and over the sea, on land (with the Royal Marines) and ever more frequently in the littoral, fully interoperable and integrated with UK and allied forces. To support this transformation we are investing in two large aircraft carriers to operate the Joint Combat Aircraft (operated by both RAF and RN personnel) and amphibious shipping. These new Fleet units will transform our carrier strike capabilities and enhance our ability to project power onto the land environment. They will be supported by our future Air Defence ships, the Type 45 destroyers and new nuclear powered attack submarines, the Astute class, which will also contribute to the securing of our maritime supply routes.

Royal Navy and Royal Marines personnel assist the local inhabitants of Kalmadu Village during Op GARRON, the UK response to the Tsunami

A2.9   Land - The operations in Iraq and Afghanistan in the last few years have demonstrated the continued need for war-fighting capabilities at the high intensity end of the military spectrum. At the same time both theatres, and in particular Iraq, revealed the essential need to have the ability to rapidly reconfigure forces in theatre as the conflict develops from heavy war-fighting to enduring peace support and post-conflict reconstruction. The ability to meet this changing requirement is recognised by Future Army Structure, which involves a shift from the current mix of light and heavy forces to a more balanced structure of light, medium and heavy forces. This change will be underpinned by some key elements including Apache (which is already in-service), Guided Multi Launch Rocket System (GMLRS) and Indirect Fire Precision Attack (IFPA) providing an agile precision attack capability, and the Future Rapid Effect System (FRES) family of vehicles which will

replace many obsolete vehicles and form the core of our future medium weight force. In parallel the effectiveness of our land forces will be further enhanced through Bowman, networked surveillance systems such as Watchkeeper and ASTOR, and improved digitization and simulation of our training. There is also a substantial equipment programme underway to improve our dismounted combat capability. These changes will allow our Army the ability to conduct short notice expeditionary operations more efficiently; allowing a more effective response to a wider range of possible contingencies, while at the same time retaining the ability for war-fighting at large scale effort with heavy forces if necessary.

A2.10   Air - Airpower has been shown to be a critically important aspect of modern warfare. The RAF is driving forward modernisation to create a flexible and agile Air Force equipped with highly capable multi-role aircraft such as Typhoon and Joint Combat Aircraft, increasingly able to exploit networked capabilities and equipped with a range of advanced stand-off precision weapons. We are purchasing the fleet of four C-17's we currently lease plus an additional aircraft, which will operate in the future alongside the new A400M military transport aircraft. ASTOR and Nimrod MRA4 will improve situational awareness for our commanders to bring them improved level of accuracy and speed of response. These changes aim to ensure that we can adapt to new threats and environments and that we are able to maintain air superiority and deploy forces worldwide in the future.

A2.11   The chapters within Part B further consider the future strategic themes in the context of specific industrial sectors. Each sector chapter looks closely at the future requirement for capability to support the Armed Forces and identify the UK industrial base that is essential in achieving these requirements.

Selection criteria

A2.12   Every nation ideally wants to keep under its control critical defence technologies, but no country outside the US can afford to have a full cradle to grave industry in every sector, and our Armed Forces continue to benefit from the extensive range of foreign-sourced equipment currently in service. And it is readily recognised that much of the equipment procured from UK prime contractors contains non UK sourced content. We welcome the progress made in establishing understandings on Security of Supply (the US/UK Declaration of Principles and the Letter of Intent Framework Agreement). Moreover, we recognise the potential benefits that may be realised from the November 2005 decision by the EU Defence Ministers to introduce a Code of Conduct on Defence Procurement which aims to create an effective European Defence Equipment Market. We continue to welcome overseas products offered in competition, and indeed in many significant areas rely on overseas supply, with appropriate guarantees (which may include technology access to ensure we can adapt equipment to meet national requirements over time) and/or judgement that any increased risk to maintaining our operational independence is acceptable.

A2.13   The UK also retains a sizable, open and broadly-based defence industry which delivers a large proportion of MOD's needs, and we welcome overseas investment, especially from companies that create value, employment, technology or intellectual assets in the UK and thus become part of the UK defence industry. Within this strategy, we aim to tell industry very clearly where, to maintain our national security and keep the sovereign ability to use our Armed Forces in the way we choose, we need particular industrial capabilities in the UK (which does not preclude them being owned or established by foreigned-owned companies). We have therefore assessed industrial capabilities selection criteria developed to assist us in identifying which aspects of the UK industrial base are essential for us to sustain onshore in order to deliver the capability the Armed Forces require. This process was trialled jointly with industry under the auspices of the National Defence Industries Council (NDIC). The criteria reflect the overarching principle of Appropriate Sovereignty - indigenous industrial capability required for retention for national security reasons, as discussed earlier in the strategy. They fall under the following headings:

- **Strategic assurance** - Capabilities which are to be retained onshore as they provide those technologies or equipment important for the safeguard of the state. Such technologies could include those used within the nuclear deterrent, high-grade cryptography or that are a key tenet of our counter terrorism capability. Industrial capabilities could be classified under this heading if for instance:

  - the technology is of such strategic importance that the risk of obtaining the required industrial capability from overseas is unacceptable. The UK Government could simply not countenance sourcing the capability from overseas;

  - we require minimal or no risk of the capability failing given its strategic importance. The risk of sourcing the capability from overseas increases the likelihood of associated military capability failure;

  - we need to prevent an adversary from acquiring knowledge which could present them with the capacity to prevent our Armed Forces effectively deploying or using the associated military capability;

  - procurement or sourcing the industrial capability from overseas is prohibited for legal reasons, e.g. contravening treaty obligations.

- **Defence capability** - The retention of equipment and technology within the UK industrial base is necessary as our Armed Forces require particular assurance of continued and consistent equipment performance. This could include those of particular operational importance, or aspects of more generic battlefield systems or sub-systems, where failure could present particular danger to our Armed Forces. Industrial capabilities could fall under this heading if, for instance:

  - there is a specific need to assure the security of supply of an industrial capability or technology. There could be a viable global and open marketplace for the industrial capability, however the assurance of supply could be suspect and would present too great a risk to the availability and reliability of effective equipment performance for our Armed Forces. In some cases, we may not yet be able to ensure sufficient access to Intellectual Property Rights from overseas to provide confidence in our ability to understand its true operational performance or to update the equipment over time, in which case there may be little option to seeking to produce an indigenous capability.

  - failing to sustain the capability in the UK would lead to reliance on a single overseas source. This may in itself present a risk of adverse political interference from the relevant overseas Government at the point that the UK seeks to procure from the relevant supplier.

  - it is important to ensure that associated Intellectual Property (IP) within the UK is protected and not transferred to a potential adversary. Transfer of such IP to an adversary could result in replication of our military capability and possible use against our own Armed Forces.

  - it is particularly important that our Armed Forces have secure <u>priority access</u> to the industrial base needed to enable effective equipment acquisition, support and upgrade when and where we require it. This may be of particular importance when seeking to meet UORs. This is closely related to the broader concept of security of supply, but applies when a foreign supplier might otherwise be willing, but may be unavailable to meet the requirement in time - either because of the distances involved, or because their domestic customer has first call on his capacity.

- **Strategic influence** - where specific UK capabilities give us important strategic influence, in military, diplomatic or industrial terms. Collaborative or complementary programmes may often be relevant here. Such programmes may be pursued to ensure value for money and affordability in complex programmes and to help enable cohesive coalition operations. The UK continues to enjoy the ability to actively participate in such programmes partly as a result of an industrial base which has a strong history of providing world-class capabilities and technologies across all military environments and platforms. This heritage gives the UK Government leverage in and access to these programmes. Once engaged, we can better ensure that our requirements can be met cost-effectively and bringing to bear the strengths of UK industry. But strategic influence can also come at the military and strategic diplomatic levels, especially when the UK can bring something distinctively different to the table. Thus capabilities prioritised under this heading may have one or more of the following characteristics:

  - continued possession and development of an industrial capability onshore enables access to a unique UK military capability. This military capability also enables UK military planners to have particular influence in the development of coalition operations, and the acknowledged merits of the capability concerned may increase diplomatic influence more generally;

  - it has particular cutting edge characteristics which ensures that the UK can ensure due weight is put on its requirements and achieve equitable and fair technology share in collaborative equipment or research programme with other nations;

  - it is important to ensure access to a particular collaboration project. This in turn may help to strengthen political and diplomatic relationships between the UK and another nations.

A2.14 **Technology Benefits** - The UK defence industrial base has in the past been a productive and innovative source of technologies and capabilities which have had follow on civil applications. Defence has then benefited from the associated economies of scale. There continue to be opportunities to pull through military technology to the civil market, recent advances in biomedical screening are a good example. Therefore, the continued investment by Government in certain areas could help enable further opportunities. We recognise that this is not always directly an issue of national security, but it is a consideration to capture, in ensuring that our policy is consistent with broader Government policy on promoting innovation. In practice, we have not found any capability which was important to retain only for this reason.

Defence planning

A2.15 In order to understand the nature of our forward plans and the way they are described in this document, it is important to recognise the framework within which they sit.

A2.16   MOD policy staff develop **Defence Strategic Guidance** (DSG), a classified document developed in consultation with other government departments and approved by the Secretary of State for Defence. The DSG includes detailed defence planning assumptions and is informed by a comprehensive methodology and suite of analytical tools. It is our principal strategic direction for the development of Defence and the policy baseline for our planning cycle. It establishes key planning parameters and priorities for resource allocation and capability development, stretching out broadly over the next 15 years.

A2.17   Government **Spending Reviews** usually take place every two years (although the last Spending Review, which concluded in 2004, will be followed by a Comprehensive Spending Review reporting in 2007). Spending Review settlements conventionally set Departmental budgets for the following three years, with the final year of the period forming the baseline for the next Spending Review. Whilst Departmental budgets may be subject to adjustments in the intervening period - for example, to respond to fiscal shocks or in the case of Defence, to reimburse the Department for the net additional cost of operations - budgets should essentially be regarded as fixed between Spending Reviews.

A2.18   We conduct an internal **planning round** to incorporate the overall Defence budget set in the Spending Review in a forward **Defence Programme** and underpinning resource allocation to budget holders. The Defence Programme is reviewed and adjusted every two yeas, and comprises two core elements: the **Short Term Plan** (STP) and the **Equipment Plan** (EP). Beyond the end of the Spending Review period these are only indictive planning assumptions, recognising that budgets can go down as well as up.

A2.19   The STP looks out over a four year period, and covers the running costs of Defence and some areas of capital investment. Aside from covering major areas of expenditure such as pay and investment in estates and business information systems, of particular interest in the context of the DIS is that the STP provides the budget for the Science Innovation and Technology organisation and the Defence

Logistic Organisation, for the support of in-service equipment.

A2.20   The EP covers a 10 year period, and sets out how we plan to spend the funds provided for the acquisition of new equipment for the Armed Forces. The requirements for new equipment are set by the Equipment Capability Customer (ECC), whilst the procurement is carried out by the Defence Procurement Agency.

A2.21   Within the ECC, it is the responsibility of the **Directors of Equipment Capability** (DECs) to develop coherent equipment capability solutions. The DECs who report to a **Joint Capabilities Board**, chair **Capability Working Groups** (CWGs) are to assess and develop options for the delivery of future capability programmes. CWGs are the structure through which the relevant stakeholders come together to co-ordinate planning across all the Defence lines of development, including organisation, concepts and doctrine, training, infrastructure, information, personnel and logistics as well as equipment, in order to deliver military, rather than simply equipment, capability.

A2.22   DECs conduct **Capability Audits**, underpinned by operational analysis, military assessment panels and balance of investment studies, to determine whether they are meeting their capability objectives. The audits use a set of planning scenarios - derived on the DSG - against which to assess whether or not the EP will deliver the capabilities UK forces need to meet the most likely operational tasks of the future, including the requirement to conduct operations of a certain scale simultaneously. The audits feed into a **Capability Area Plan** (CAP) which is an authoritative statement of a DECs capability surpluses and shortfalls, and a route map for future capability development. The CAP guides the activity of the CWG and is the source of any proposals to adjust the EP.

A2.23   It is important to understand what this means in practice about the nature of Defence planning and our forward plans:

● Our formal budgets only extend, at most, three years ahead. Where the STP and the EP extend out beyond the time horizon of the government's formal spending plans set in the Spending review, these are internal MOD planning assumptions and are subject to change as the government periodically reassesses its priorities. Whilst it is essential that we plan beyond the period for which we have a formal Defence budget from the Spending Review, planning assumptions and costs are inevitably ever more uncertain the further out into the STP and EP period one looks.

● Whilst the Treasury will take existing Departmental plans into account as we set new budgets in the Spending Review, no Department - including the MOD - can assume that future settlements will continue past trends in budgetary allocation.

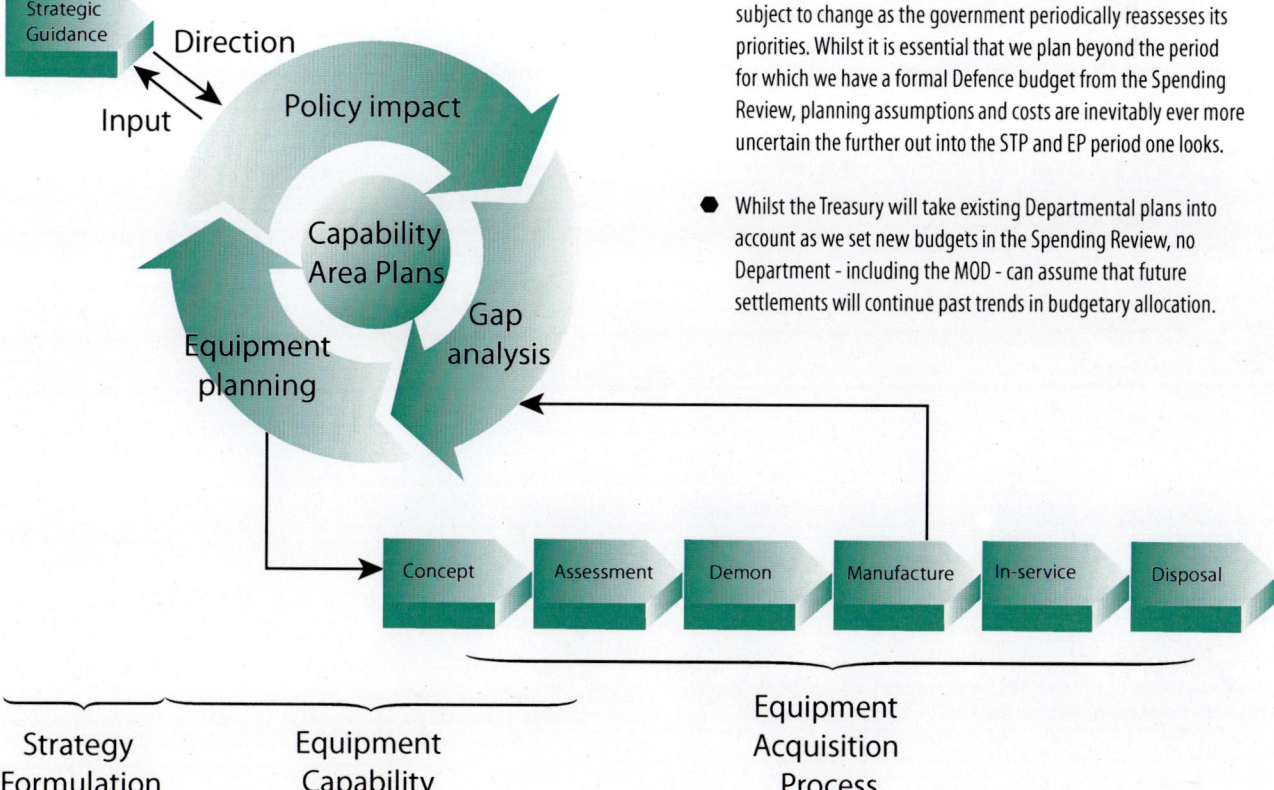

- Within a budget set by the Spending Review settlement, our planning is essentially a process of prioritisation. Decisions to allocate more resources to areas of high priority must be offset by savings elsewhere.

- It is very important to maintain flexibility in our planning. The security challenges we face, and hence Defence priorities, will always be subject to change, as will cost. Our planning must be flexible enough to respond to such changes in the strategic and resource environment.

- Conversely, our ability to make adjustments to the forward Defence Programme in the shorter term is more constrained. For example, as figure A2(ii) illustrates, the high level of contractual commitment to existing projects in early years of the EP limits (although does not preclude) our ability to make changes in this period.

- For a project to have an EP or STP funding line does not necessary mean that the project has been given approval to proceed to the delivery phase. It is only at Main Gate, the main investment decision point, that the commitment to the solution and specific performance, cost and time parameters is made.

- Looking to the longer term, we will tend (save for where specific contractual commitments have been made) only to have an intent or capability aspirations beyond the EP period, together with a view of strategic priorities and the broad order financial consequences of the most significant programmes. Knowing the industrial capabilities likely to be required to meet such aspirations will be dependent on technological developments and investigation of the full range of potential solutions to the capability need.

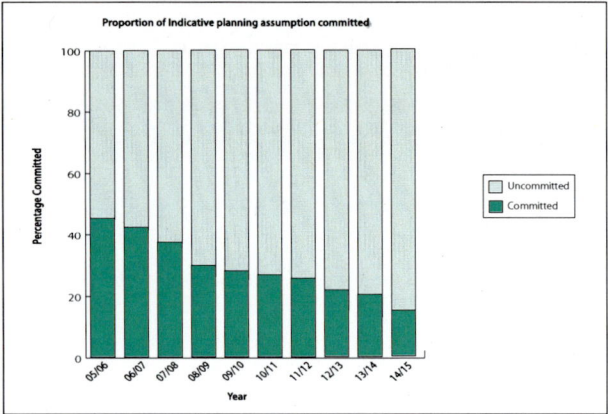

Figure A2(ii).

Royal Navy Sea King land Royal Marines from 42 Commando, during Operation Silkman, Sierra Leone.

The Defence market

Introduction

A3.1   This chapter outlines the main characteristics of the defence market at global and national UK levels, identifying recent changes, current state and forecast near term trends. This understanding provides important context for assessing the likely evolution of our supplier base, the implications and the degree to which we and wider Government can assume levels of positive influence and where necessary control over the supplier base to secure DIS objectives. The DIS does not seek to set out a preferred route to international restructuring; that is very much industry's business. But it does seek to create a clear UK context to inform these decisions.

A3.2   The chapter finishes with a framework of levers available to governments which affect industry at both the general level, i.e. the overall attractiveness of the defence business environment in a particular country, and at the specific level, to achieve defined outcomes in particular capability or technology areas. The next chapter considers in more detail how these are applied in the UK.

Global perspective

A3.3   The last 15 years have seen major changes in governments' defence requirements and funding profiles. Driven by the demise of the Warsaw Pact, global defence spending fell sharply in the 1990s by a third in real terms from $1,300 billion (or around £800 billion) in 1989 to $800 billion (or around £500 billion) in 1996[1] . In the US, the world's largest defence market, the cuts were particularly significant and had a major impact on industry, driving rationalisation and consolidation within the US supply base. A similar, but less pronounced, effect was seen in Europe.

A3.4   However, since 9/11 the emergence of new threats and the emphasis on national security has resulted in increased defence spending, most notably in the USA, with increases, albeit not on the same scale, also in most European countries, including the UK and France. Following an 18% increase in real terms over the last few years, global defence spending has now reached nearly $1,000 billion (or around £600 billion)[2].

A3.5   Of this, the defence market (i.e. Governmental spend with industry) is estimated to be worth almost £200 billion worldwide. This is largely split into discrete national markets. As figure A3(i) shows, the USA is by far the largest market at around £90 billion in 2004. The implication of this scale, and the high proportion spent on research and technology, is that the DoD leads in the development and exploitation of key enabling defence technologies and tends to have an influential lead in developing advanced military doctrine and concepts for use of those technologies in an operational environment.

A3.6   At the next level down, despite Russia, China, Japan, the UK and France all being major defence spenders, the UK market is significantly larger than the others because we are increasingly relying on industry to take on new roles beyond equipment development and manufacture; an increasing proportion of our budget is spent sourcing products and services from a largely private sector industrial base.

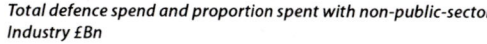

Total defence spend and proportion spent with non-public-sector Industry £Bn

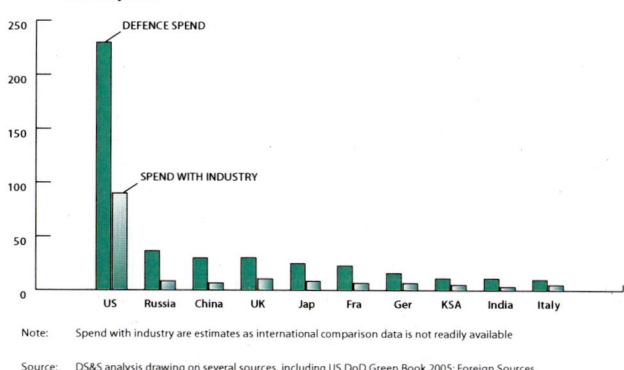

Note: Spend with industry are estimates as international comparison data is not readily available

Source: DS&S analysis drawing on several sources, including US DoD Green Book 2005; Foreign Sources of Supply, Assessment of the US Industrial base 2004; DASA statistics; Centre for Arms Control and Non-Proliferation 2005, Military Balance 2005.

Figure A3(i).

A3.7   Not only is the USA the largest spender, it has also continued to increase spend with a rise of 4% in real terms since 2002. The largest growing category of spend is research and development which has grown 11% in the same period[3]. With these disproportionate levels of defence spending, the defence industrial, technological and military gap between the USA and the rest of the world continues to grow. Indeed, in some areas, the USA aspires to have technology 'way ahead' of others[4].

Industrial development

A3.8   In response to the falling defence budgets of the 1990s, significant consolidation took place in the defence industries of the USA and Europe.

A3.9   In the USA the consolidation was framed by the government's commitment to rationalisation whilst retaining competition at the prime contractor level. This resulted in the creation of five large and globally important US defence companies: Lockheed Martin, Northrop Grumman, Raytheon, Boeing and General Dynamics.

A3.10   European consolidation progressed more slowly, in part due to the continuing political pressure to retain capability at a national level. European rationalisation took a number of forms: internal national mergers (e.g. British Aerospace with GEC), the use of joint ventures to bridge national boundaries (e.g. MBDA) and limited cross-boundary consolidation (notably EADS). As a result there are now several large European companies – namely BAE Systems, EADS, Thales and Finmeccanica. Further rationalisation is still expected. Within the UK consolidation has been taken further than in wider Europe and the industrial structure is now relatively mature and stable, although further rationalisation within this construct is possible.

A3.11   US companies continue to dominate the global defence industry. As Figure A3(ii) below demonstrates, 7 out of the top 10 defence companies are now US-based. BAE Systems is the fourth largest international defence company reflecting not only its leading UK position but also success in accessing a share of the large US market. Whilst these defence companies are giants within the defence market, accounting for around 50% of global sales in 2004, they are relatively small in global corporate

---

[1] *International Institute for Strategic Studies Military Balance*
[2] *Stockholm International Peace Research Institute*

[3] *DoD Green Book 2005*
[4] *Defense Industrial Base Capabilities Study*

investment terms. For example, Lockheed Martin ranked 135th in the Global 500 on 2004 revenues whilst BAE Systems ranked 399th.

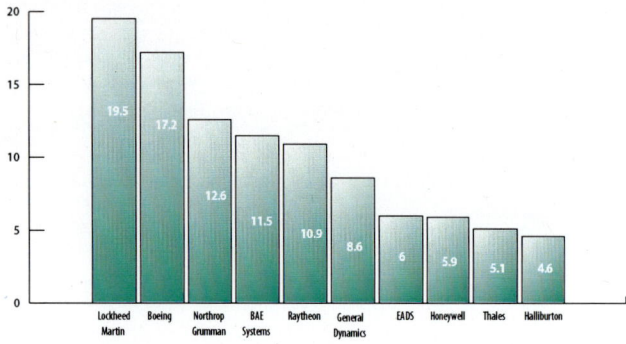

Global top 10 defence companies based on defence revenues 2004 (£Bn)

Source: Defence News, 2005

Figure A3(ii).

A3.12   Not all of these companies have the breadth and scale to deliver the broadest range of defence business, at the integrated platform and system level, across the land, sea and air environments. These include, for example, BAE Systems and Lockheed Martin. Others such as Raytheon and Honeywell, whilst still large defence players, are more focused on specific types of business.

A3.13   The major defence companies, who take on the delivery of complex, integrated systems and platforms, are heavily dependent on a wide range of lower level suppliers, many of whom are significant international companies in their own right. A large proportion of the specialist intellectual property and innovation so essential to delivering world-class military capability is believed to lie in these lower tiers of the supply base.

Cooperative programmes

Cooperative programmes, undertaken with allies whether in the USA, Europe or elsewhere, can bring together governments and industry and sometimes act as the stimulus for industrial restructuring. They also offer potential benefits in their own right:

- defence research cooperation offers economic and technology benefits, with an assessed 5:1 return on investment and providing knowledge with an annual value of around £280 million through information exchange programmes;

- economic benefits through sharing development costs and through economies of scale in production and sharing in-service support and upgrade costs;

- enhanced interoperability with allies;

- jstrengthening bilateral relationships, including security relationships.

It is important, in cooperative programmes, that we still retain enough understanding of the underlying intellectual property, including from our partners, to be able to adapt our equipment through-life to meet national requirements, as discussed at the end of chapter A2.

European perspective

A3.14   The European defence environment can be characterised as a set of largely separate domestic markets. In general continental European markets are both smaller than the UK market and less open.

A3.15   The desire to make European military contributions more effective, combined with the economic realities associated with nations sustaining largely separate markets, has nevertheless prompted initiatives in cooperative procurement including several major programmes (such as the Airbus A400M and MBDA Meteor) utilising OCCAR[5] and more recently the creation of the European Defence Agency. A recent Commission Green Paper on the European defence market[6] recognises that greater efficiencies can be achieved in European defence procurement. However, as in other European policy areas, there are significant integration challenges and national interests remain a dominant factor. EU enlargement adds a further dimension of complexity since countries such as Poland are highly aligned with the USA and will often favour purchasing US defence products and services. Enthusiasm from governments and industry for collaborative development programmes has undoubtedly been tempered by difficulties experienced on some high profile collaborative ventures in recent years.

A3.16   In practice, the European market remains fragmented although it is hoped that the European Defence Agency will begin to make a difference in terms of supporting the more effective harmonisation of military requirements and promoting a more open defence equipment market. Progress is being made in opening project procurements to European competition and addressing security of supply concerns, as outlined at A3.32 below. However, at current spending levels the market cannot offer the same scale and scope as the US market. Furthermore, European national defence markets are expected to grow at lower rates allowing the US market to continue to pull ahead.

A3.17   Historically European governments owned and controlled much of their indigenous supply base. In the post Cold War period there has been a trend towards privatisation or partial privatisation of previously state-owned enterprises (e.g.MTU in Germany). This has resulted in a shift of emphasis towards achievement of shareholder value rather than the delicate balance previously sought between industrial performance and national ambition and, as a result, is attracting interest from private equity investors. Affordability constraints however are likely to force governments and companies to tackle overcapacity of design and production in some sectors.

A3.18   Continental European companies, as well as securing their position in domestic markets, are trying to access other markets and increase exports to compensate for generally reducing or constant domestic budgets. The UK defence budget has grown, but UK companies still generally are seeking to secure a share of the larger and generally more profitable US market. UK companies continue to invest in the USA, making a total of around £2 billion of US acquisitions in 2004 alone in almost forty separate acquisitions. British companies such as BAE Systems, Rolls-Royce, Smiths Group, VT and QinetiQ have bought US companies to overcome the high entry barriers and secure progressive access to the market. However, a continuing commitment to the UK market combined with the constraints on accessing and operating in the US market, forces difficult boardroom decisions for UK companies on where to locate core capability and investment.

---

[5] OCCAR (Organisation Conjointe de Coopération en Matiere d'Armament) is the European procurement organisation which manages a number of joint equipment programmes such as the A400M military transport aircraft and the Principle Anti-Air Missile System (PAAMS) which will be used on the UK's Type 45 Anti Air Warfare Destroyer.

[6] Towards an EU Defence Equipment Policy" – Com(2003)113 dated 11.3.2003

A3.19   This trend will continue as long as the US market is disproportionately attractive in scale and relatively closed in comparison to other markets and as long as entry costs remain justifiable from a shareholder value perspective. Exports to the USA at the platform level are possible, as demonstrated by the recent sale of the Agusta Westland US101 for the Presidential flight, but a high level of US content is required. The financial markets clearly currently view the USA as delivering the best opportunities for shareholder value.

A3.20   The key question is whether investment into the USA is coming at the expense of further investment in the UK market. The political environment in the USA and recent trends do not suggest that technology sharing will get significantly easier in the near term, and the ease of sharing technology across national boundaries can affect significantly where companies choose to invest, particularly in R&D. However, these problems, as well as cost competitiveness, can also lead US companies to invest directly in the UK. For instance, Raytheon Systems Ltd, a wholly owned subsidiary of the US Raytheon Company, now employs 1500 people across the UK and, as well as engagement in a number of UK projects, is a net exporter, including back to the USA. DTI estimates that at present defence investment into the UK is exceeded by outward investment from the UK into the US, but it is unclear whether this is likely to be an ongoing trend, as the most attractive acquisition opportunities are taken and as companies seek to integrate their recent acquisitions.

A3.21   Companies based in continental Europe also look to achieve access to other markets and improve their export success. Whilst UK companies have been relatively successful in establishing trusted supplier credentials in the USA, other European companies have had less success. Instead their attention has been focused on accessing the UK market which is closer to home, relatively open to foreign suppliers and shareholders and an attractive extension from their home markets. Thales and Finmeccanica are examples of foreign-owned companies who have successfully established significant UK market share, generally by acquisitions. Given the close defence relationship with the US but also a central role in several pan-European defence equipment programmes, the UK, as a base, can offer a bridge between Europe and the US.

UK perspective

A3.22   The UK defence market remains a considerable size in its own right, as we will consider further below, and the Defence budget has benefited from the longest period of sustained real growth in the UK's defence spending plans for over 20 years. In real terms, defence spending will be 7.5% higher than 1997/98 by 2007/08. In addition, the UK defence market has some defining characteristics compared with other relevant national markets in terms of:

● sophistication of demand;
● market openness and diversity of supply;
● profit potential and the trading environment.

A3.23   These characteristics define its relative attractiveness to industry, its ability to attract investment and the level of influence or control UK Government can be confident of achieving in the industrial base. Each aspect is discussed below highlighting the current situation and identifying change drivers and trends.

Sophistication of demand

A3.24   We buy a wide range of defence products and services: from basic items to complex integrated systems; from one-off purchases to long term support services. The UK aims to maintain a capability edge and must maintain adequate interoperability with US equipments, particularly command and control systems. To achieve this we invest in research and development to ensure we can remain at the forefront of important defence innovations. As such the UK provides an attractive

home market for product development and subsequently a sound platform for exports. Our indigenous industry is broadly-based, covering all environments and including civil-based information and communications technologies (discussed further below). It also includes enabling capabilities that add significantly to effectiveness and value, e.g.: propulsion, radars, power generation and management, platform signature management, synthetic environments and training, electronic warfare algorithms, and open architecture systems. Many of these enabling subsystems are promoted and exported in their own right and have been selected on merit by prime contractors in the USA and elsewhere.

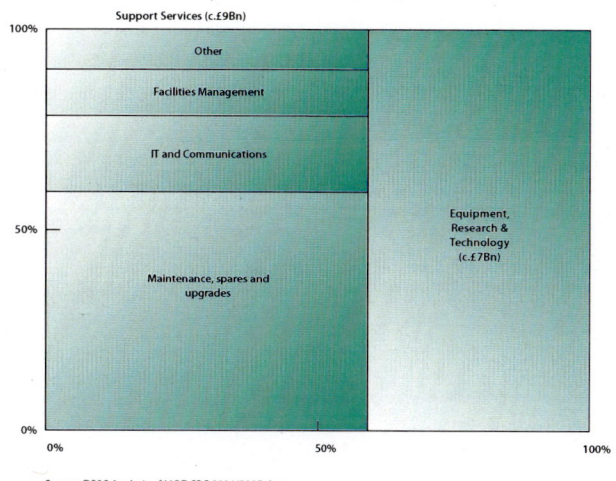

Estimated Market Segment Sizes (2004/05) £Bn. Total = c.£16Bn

Source: DS&S Analysis of MOD SRG 2004/2005 data

Figure A3(iii).

A3.25   Whilst platform-related activity still shapes the UK defence market, the figure A3(iii) shows that we already spend up to £9 billion on support contracts with industry, covering logistics, maintenance, repair and upgrade, IT enablers and facilities management; this represents a very significant proportion of overall external spend.

A3.26   **Through life management** - The spending emphasis on support services is likely to persist. Our commitment to effective through-life management of defence capabilities and assets to improve capability and agility, enable technology insertion and reduce whole-life costs creates opportunity for industry. Our Defence Logistics Organisation and its key suppliers are already establishing innovative arrangements around key programmes to move industry's support role away solely from supply of spares and maintenance services towards supply of asset availability and incremental upgrade of capability. Similarly future programmes, such as the Future Strategic Tanker Aircraft, are being defined to address through-life support as part of the initial acquisition process. This trend is set to continue, with many of the processes and roles currently undertaken by the MOD likely to be delivered in future through partnership with industry. This requires the development of acquisition models that engage a range of industrial players including equipment design authorities (e.g. aircraft Original Equipment Manufacturers), technology inserters (such as defence electronics companies), integrators of complex systems and/or military capability integrators and innovators (e.g. niche technology companies).

A3.27   **Focusing on Military Capability** – The military strategic overview chapter outlined the emphasis on effects-based operations and the need to plan and manage the defence business at the level of "military capability" i.e. integrated and agile combinations of people (appropriately trained and supported), equipments (appropriately updated and maintained), infrastructure and information, and structures and processes that can create military effect in a range of operational scenarios. This higher level approach provides opportunities for us and industry to

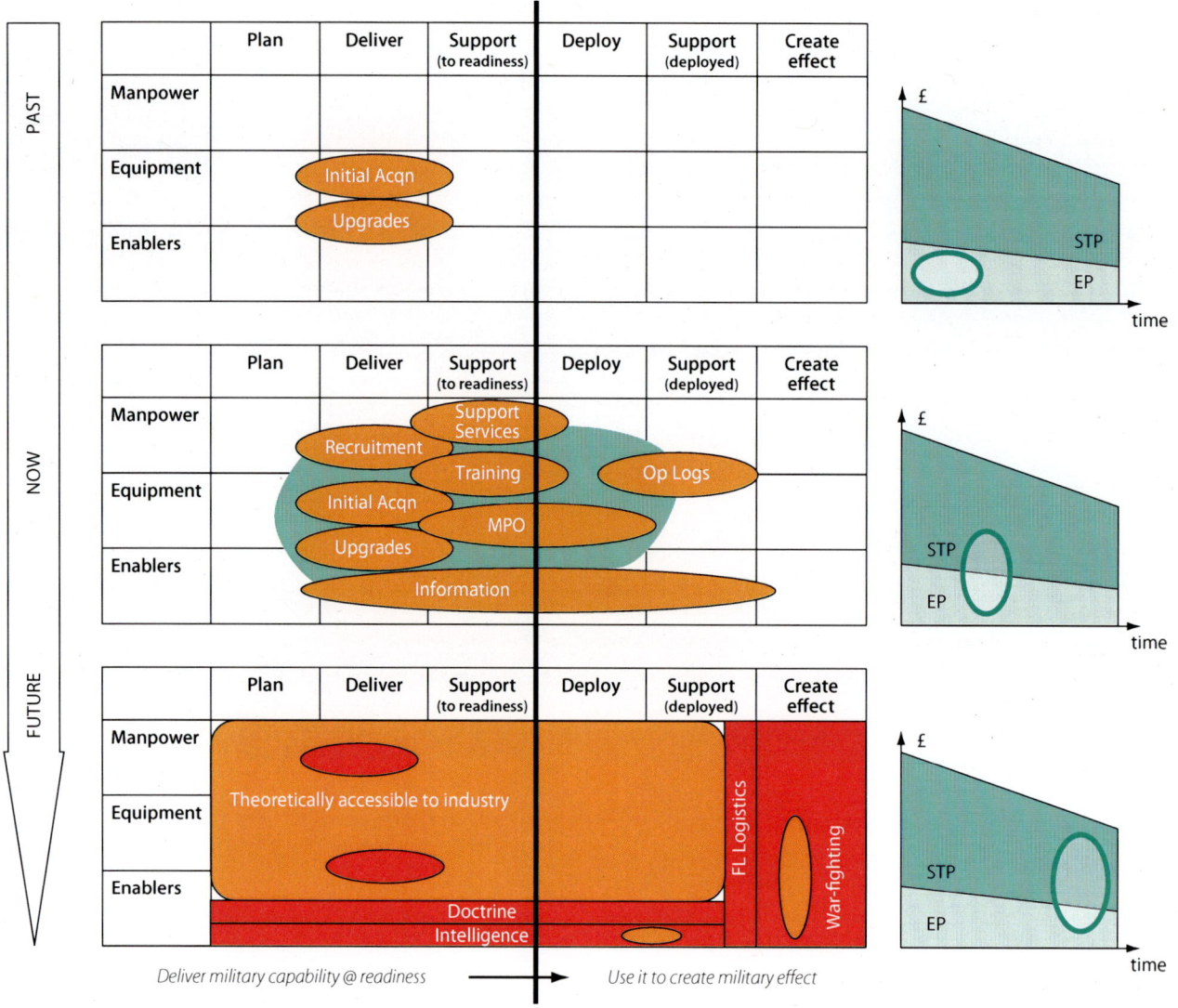

Figure A3(iv).

think more innovatively about the allocation of roles and responsibilities and the use of commercial models to make best use of our and industry's comparative advantages in delivering and supporting military capability.

A3.28 **Market implications** - Figure A3(iv) above shows that in practice this intent implies a continuation of the rebalancing of roles that is already underway:

- In the past, industry's role (shown in orange) was limited to the provision of our equipments, upgrades and equipment-focused support services. Our relationships with industry were transactional. Industry's share of the defence budget was focused in the Equipment Programme;

- Now we are more explicitly dependent on industry for a wider range of products and services. Industry roles extend into non-equipment areas and already span the boundary between peacetime and deployed environments. Products and services previously supplied separately are now being grouped into larger integrated packages (including through-life, system-of-systems and cross-platform) and industry now accesses the Equipment Programme and a substantial portion of the Short Term Plan (STP);

- In the future, assuming this trend continues, an increasing

amount of the defence budget would be made accessible to industry and the packaging up of different elements of military capability at the MOD/industry interface would continue. We would retain core roles (shown in red) but become more explicitly dependent on key suppliers for delivery of defence outcomes. Any individual changes to the MOD/industry boundary are, of course, likely to require consultation with Trade Unions and demonstration of better value against a public sector comparator, but the overall trend over the last few years is recognisable.

A3.29 **Increasing importance of Information & Communications Technologies (ICT):** Whereas the differentiator in military operations could previously be measured in terms of scale and potency, today it is more about agility and the ability to create an appropriate military effect rapidly and in response to specific information about the operational environment. Whereas technologies such as stealth, advanced signal processing and energetic materials have generally been defence-led and largely under the control of defence sectors, it is the commercial sector that is driving innovation in ICT. Last year US businesses invested more than £40 billion on ICT research and development alone, more than the entire US Defence R&D budget[7]. To remain at the leading edge of military capability will, therefore, increasingly require effective exploitation of commercially-driven ICT.

A3.30 Civil ICT is characterised by open international standards, fast technology cycle-times and high levels of investment from the major global players. A range of commercial supply models is used, normally based around high volume service provision incorporating upgrades.

[7] OECD

Governments need to work with commercial developers and align with their business models and development cycles if they are to access, exploit or control the use of ICT technologies for defence purposes. This requires the use of novel approaches, for example: activity at the Government level to make the market attractive; commitment to COTS, common standards and open architectures wherever possible; and encouraging the core defence supply base to develop its ability to access new technology and quickly assess and apply it in the defence environment.

Sophistication of demand – key implications

Our demands on the industrial base are becoming more sophisticated. We see an increasingly important role for industry in delivering cost-effective military capability, managed on a through-life basis; this provides significant opportunity for existing and new suppliers in the market.

We recognise the importance of accessing commercially-led technology developments through engagement with the broadest supply base, including companies who do not traditionally specialise in defence. The DIS needs to support a business environment that develops these relationships.

Market openness and diversity of supply

A3.31   As figure A3(v) illustrates, UK defence is a comparatively open market. In 2004/05 some 5% of our spend with industry was directed at imports, a further 14% with foreign-owned UK-based companies and a significant further proportion (13%) to cooperative programmes run through European organisations such as NETMA and EUROPAAMS. Furthermore, a proportion of our spend attributed at this level flows through to increasingly international supply chains. In contrast, the USA spent less than 2% on imports and 7% with foreign-owned companies, much of which was with BAE Systems North America.

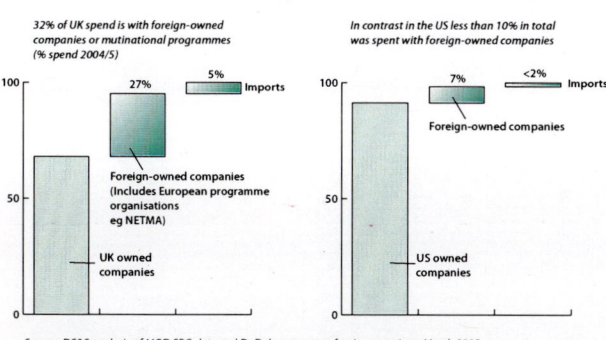

Source: DS&S analysis of MOD SRG data and DoD document on foreign sourcing - March 2005

Figure A3(v).

A3.32   The principal continental European markets remain less open than the UK in terms of foreign access to domestic markets, rules on foreign inward investment into local companies and significant retained shareholdings by some governments. However, we are encouraging others to be similarly open, and the European Defence Agency (EDA) is working to create an effective European Defence Equipment Market (EDEM), in other words an open, competitive and transparent environment that will strengthen the European defence technological and industrial base such that our Armed Forces will be able to secure their equipment capability needs more cost effectively. The voluntary Code of Conduct on Defence Procurement[8] which applies to those defence equipment procurements exempted from the EC public procurement rules on national security grounds[9] is a significant step in this direction, supporting long held UK policy aims for more open defence equipment markets achieved through a self-regulatory approach, and offering UK industry the potential for a level playing field for defence companies competing for business within the EU.

A3.33   Continental European companies have taken opportunities in the UK in their pursuit of alternative sources of growth to compensate for the lack of scale and growth in their domestic markets. For example the Finmeccanica group now owns significant parts of the UK's defence supply base, including helicopter capability in Agusta Westland and the avionics/electronics capability in Selex.

A3.34   Large US companies participate more selectively in the UK market seeing it as an optional extension to their activities in the USA. Their primary focus and priority remains with the US market and DoD. However, they are able to leverage the scale of their domestic production capabilities to compete favourably in the UK market. The financial and administrative barriers to US companies entering the UK are lower than the costs associated with UK companies attempting to build positions in the USA because of the different approaches required to establish entry. Generally, European companies have to acquire capability in the USA to establish themselves in the US market; US companies tend to be able to bring sufficient capability into the UK to be considered a UK-based supplier.

A3.35   Figure A3(vi) provides a snapshot of our main suppliers by listing the 10 largest direct suppliers in 2004/05.

A3.36   This chart only shows part of the picture. It is based on billing data and therefore only reflects our highest level direct spend with suppliers and not the flow-down of that funding into the lower levels of the supply base, which will often be international. It excludes all spend on the nuclear programme and spend with major Government-owned suppliers such as Dstl.

A3.37   The chart shows a snapshot in time and our suppliers change with major programmes. For example General Dynamics UK's prominence in 2004/05 was highly dependent on the Bowman programme. Looking forward, the profile is likely to change with new programmes suggesting enhanced roles for a number of foreign-owned companies, including EADS (Skynet 5, Future Strategic Tanker Aircraft), EDS (Defence Information Infrastructure), Raytheon (ASTOR) and Thales (Watchkeeper).

[8] agreed in November 2005 (to come into effect on 1 July 2006), Although any Member State may opt out from the provisions of the Code if they wish.
[9] These are the majority of defence equipment procurements; data collected from Member States by the EDA indicates that in 2004 around 80% of such procurements (by value) were exempted from the EU public procurement rules on natioal security grounds.

MOD's top 10 direct suppliers in 2004/5 showing location of corporate HQ

Figure A3(vi).

A3.38   Despite these caveats, it is clear that the UK has a diverse supply base even at the highest level. Several of the key suppliers are under foreign control at Group level or have a significant element of international ownership or control. The DTI estimates that around 25% of the UK's defence industrial base is foreign owned[10] . Furthermore several of our biggest suppliers are more heavily focused on non-defence or international business than on UK defence business.

A3.39   The level of influence we can expect to have with companies is determined by the importance of the UK defence market to the company's revenue and profit lines and the importance of engagement in the UK defence market as a means to access and develop new technology and capability.

A3.40   As our supply base becomes increasingly international, our influence is, other things being equal, likely to decrease. Foreign-owned companies operating in the UK market are generally less critically dependent on us as a customer than on their home customer. As a result we, although an important customer, have comparatively limited influence over these companies and in particular over strategic

---
[10] Estimate based on the number of UK-based employees undertaking defence-related work

decisions around capability development and investment, especially where other governments maintain stakes in these companies.

A3.41   For those major UK defence companies who are developing significant business overseas, notably in the US, our relative importance is decreasing, although we are likely to remain significant whilst the UK is seen as the principal home market and we are seen as a major customer.

A3.42   Another factor affecting the level of our influence is the degree to which companies are focused on defence or civil markets. The major prime contractors are generally, by nature, defence companies and must remain focused on defence markets. They may have adjacent businesses in the civil sector, for example defence and civil aerospace interests, but these are largely driven by different business models. Further down the supply chain there is more scope for leveraging common technology and capability between civil and defence sides of the business and some businesses have highly diversified and often international portfolios. In these cases our direct influence is more limited and we need to ensure that the UK, and the UK defence market specifically, is sufficiently attractive to these companies as a place to do business.

A3.43   The degree to which we procure from companies with interests beyond UK defence varies sector by sector. Across the environments MOD is dependent on companies of all types and sizes. Whilst our focus is often drawn to the large companies because of their overall importance in terms of scale and breadth and their visibility, we have to understand the role played by smaller companies, whether as part of a supply chain or as niche players. Some of these companies have specialist capabilities that are critical to our ability to act responsively and flexibly, for example in response to Urgent Operational Requirements. Whereas small companies in supply chains tend to have their primary relationship with their industrial clients, some of the specialist companies also have direct relationships with MOD. There is often a high degree of interdependency, with MOD dependent on a single source for a particular capability and the company highly dependent on MOD as a major customer. We should therefore be able to work effectively with these companies to identify and secure core capability.

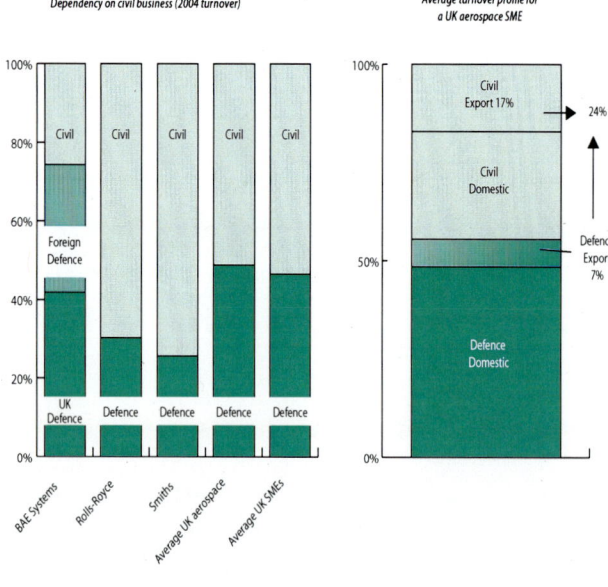

Source: Based on SBAC Aerospace Industry Survey 2005 and Company Analysis

Figure A3 (vii).

3.44   In the air environment, as figure A3(vii) suggests, key subsystems suppliers such as Rolls-Royce and Smiths Aerospace have thriving civil aerospace businesses. Small and medium sized suppliers also have strong civil business interests and a good export track record.

Market analysis and diversity of supply – key implications

We are dependent on a diverse supply base. Some suppliers are focused on UK defence and highly dependent on us. With these companies we can expect to have a high level of influence. Many others have diversified businesses, operating in the UK and overseas, in defence and civil markets; we will have less direct influence with these companies but can encourage them to participate in the UK defence market using a range of levers identified later in this section.

A3.45 In the maritime environment, the equipment and support primes are generally defence-dependent while the sub-system and component supply base is highly diversified. The anticipated spike in the new-build workload, due to programmes including Type 45 and CVF, will pose considerable challenges for the prime contractors, while the extended supply base is expected to cope reasonably well with changes in anticipated work levels.

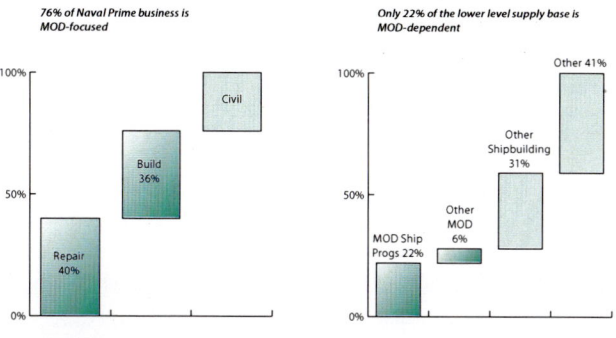

76% of Naval Prime business is MOD-focused

Only 22% of the lower level supply base is MOD-dependent

Source: SBAC, RAND Corporation 2004. *The United Kingdom's Naval Industrial Base*

Figure A3 (viii).

A3.46 In the Land environment BAE Systems Land & Armaments, ABRO and Thales are the leading defence-focused suppliers. The extended supply base is composed of companies with strong commercial vehicle businesses, as depicted in figure A3(ix).

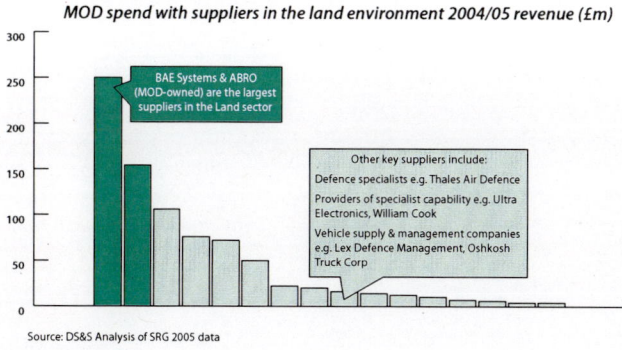

MOD spend with suppliers in the land environment 2004/05 revenue (£m)

BAE Systems & ABRO (MOD-owned) are the largest suppliers in the Land sector

Other key suppliers include:
Defence specialists e.g. Thales Air Defence
Providers of specialist capability e.g. Ultra Electronics, William Cook
Vehicle supply & management companies e.g. Lex Defence Management, Oshkosh Truck Corp

Source: DS&S Analysis of SRG 2005 data

Figure A3(ix).

Profit potential and the trading environment

A3.47 Corporate performance is directly affected by available profit margins, the selection of business models and associated contracting mechanisms, the risk/reward profile achievable within these and the approach of Government to other aspects of defence business development, such as R&D and exports.

A3.48 **Profit formulae.** We use a formula to calculate acceptable profit levels for non-competed work based on the principle of comparability with other relevant sectors within the UK economy. The US DoD takes a different approach to industry fixed-profit formulae. High risk programmes

such as technology initiatives and high risk contracts, including fixed price contracts, benefit from significantly higher margins. The approach across the rest of Europe is however largely comparable to that of the UK.

A3.49 **Acquisition models.** The selection of acquisition models may also have significancat influence. The traditional approach in the UK has tended towards tight definition of the scope of work, the use of competition to select suppliers, negotiation targeted at reducing our risk and cost, and then a transactional approach to management of the contract, holding suppliers to account against agreed milestones. More recently we have recognised that a "one-size-fits-all" approach to engagement with our key suppliers is not optimal and have deployed a wider range of supply models. The principles of partnering are now in general well understood and deployed successfully in some areas to provide mutual benefit to us and our suppliers. In the Defence Procurement Agency, several new contractual models are being deployed on significant programmes, including the use of alliances and lead systems integrators.

A3.50 **Risk/reward profiles.** Both sides recognise the need for work to optimise the approach to risk and reward. Currently industry perceives high levels of risk in our acquisition business, though their own project performance against firm or fixed price contracts is also a key determinant of profitability. Long timescales mean that a whole range of factors change during the lifetime of a programme or even a decision-making cycle, introducing risk including cancellation, requirements change, funding changes and delays. Partnering relationships, designed for mutual benefit, that recognise that risk is shared and reward performance, are more attractive to industry. The UK has a history of leading the way in deploying innovative acquisition and financing models in defence, for example PFI and PPP, and continuing to develop experience in these areas will allow us and our key suppliers to provide us with better value capability, more consistency and clarity and hence less risk, and better profit returns to industry.

A3.51 **Approach to R&D.** A government's approach to defence R&D is a critical factor determining the risk profile for industry. UK spend on R&D is significant in global terms as shown below, although US spending continues to dwarf even the combined resources of Europe.

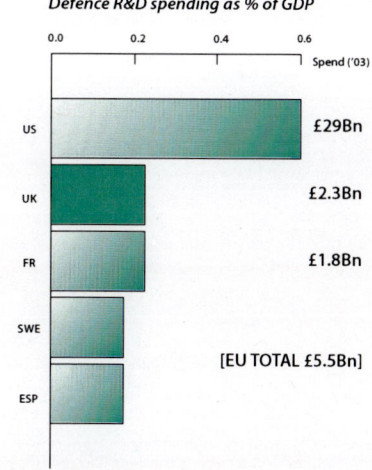

Defence R&D spending as % of GDP

Spend ('03)

US £29Bn
UK £2.3Bn
FR £1.8Bn
SWE
ESP

[EU TOTAL £5.5Bn]

Source: OECD Science, Technology and Industry Scoreboard 2005

Figure A3(x).

A3.52 In 2003, the USA invested £29billion (0.6% GDP) in Defence R&D compared with the UK's £2.3billion (0.2% GDP). The US R&D budget was 5 times that of EU countries combined (£5.5billion). This reflects a US approach to reducing risk in the early stages of defence capability development that is very attractive to defence companies and investors. Whereas the UK and other European nations often insist on securing firm price contracts with suppliers, even during early phases of programmes, the US tendency is to use cost-plus arrangements until the risk is quantified and manageable. Beyond the use of R&D to de-risk and mature technology associated with specific programmes, funding is also used to drive innovation. Chapter A10 addresses this in more detail.

A3.53 **Exports**. Exports are an important part of the defence economic landscape, offering higher profit margins than domestic sales (because development costs are already amortised) and, therefore, reducing the overall cost of a nation's defence capability procured from its indigenous suppliers. UK companies continue to consolidate Britain's position as the second largest defence exporter, with a 20% share of the global market, despite reduced defence spending in many countries and increasing competition from traditional and newer supplier nations.

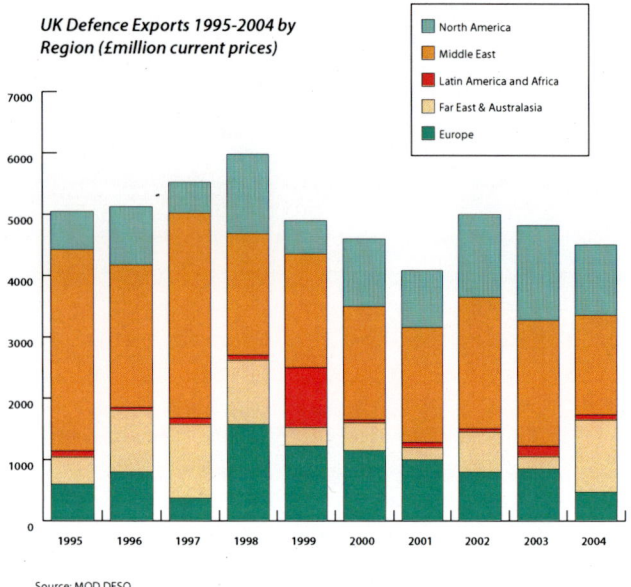

UK Defence Exports 1995-2004 by Region (£million current prices)

Legend:
- North America
- Middle East
- Latin America and Africa
- Far East & Australasia
- Europe

Source: MOD DESO

Figure A3(xi).

A3.54 All of these factors contribute to profitability. US companies tend to be consistently more profitable than European equivalents.

Profit potential and the trading environment – key implications

A number of factors drive profitability for companies engaged in our business, including uneven project performance. The DIS provides an opportunity to create a more attractive trading environment that enables industry to maximise profits whilst still retaining the focus on delivery of cost-effective capability to us.

Levers Government can use to shape the business environment for defence companies and investors

A3.55 As earlier sections have described, we need certain features in the industrial base, namely:

- a market that is sufficiently attractive to retain companies' participation against the pull of other foreign or adjacent markets;
- industrial delivery: effectiveness and efficiency;
- innovation and technology/capability development in industry;
- sufficient control for Government to ensure appropriate sovereignty.

A3.56 Governments have a range of levers with which to respond, and these take many forms. Some are available through day-to-day activities, for example as a customer of defence products and services (though here too, other government departments may have a role as customers, such as for equipment required for both defence and civil emergency services, or for secure communications). Others come from governments' wider investments in the well-being of industry and some through governments' authority as a controller or regulator (including potentially a de-regulator) of industry.

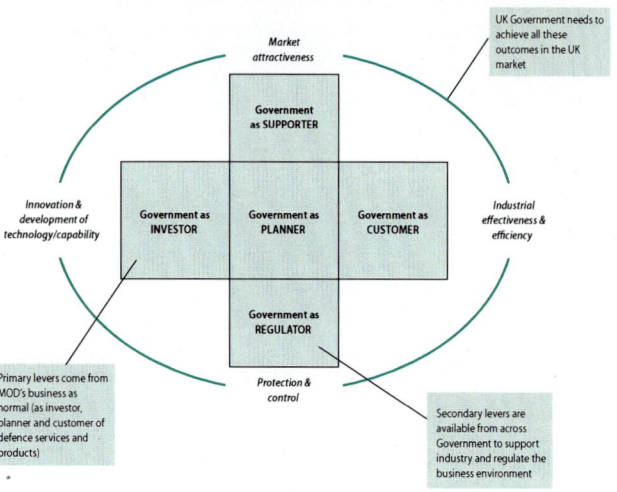

Figure A3(xii).

A3.57 Figure A3(xii) shows five groupings of levers that can be used to achieve required defence industrial outcomes. There are three groups of levers available to governments through their execution of day-to-day business, namely:

A3.58 **Government as Investor** – Whilst much of governmental investment in research and technology exists to support intelligent customer status, some is also intended to secure appropriate technology and innovation in the industrial base. Careful deployment of a range of investment levers can support technology development in critical parts of industry. Furthermore, governments' approaches to R&D are an important factor in determining the overall attractiveness of the defence market because it impacts on risk and profitability.

A3.59 **Government as Planner** – governments' forward planning activity, at the strategic and business level and addressing military capability, equipment investment and business operations can provide industry with a level of future market understanding that underpins business strategy and corporate investment. Joint planning approaches take this a stage further, providing planned alignment and interdependency between us and our key suppliers.

A3.60 **Government as Customer** – The approach taken to acquiring defence products and services - from the acquisition model chosen to the selection of suppliers and the profit margins available - fundamentally defines the attractiveness of the defence market.

A3.61 Beyond these levers, there are others available to governments in two general categories:

A3.62 **Government as Supporter of industry** – A range of levers, both financial and activity-based, are available to provide support in a number of forms to industry. In the UK, ownership of these levers is vested across Government, notably in the MOD, DTI and HM Treasury, and also in the UK's Regional Development Agencies and Devolved Administrations.

There are some key principles the Government will deploy in selecting and deploying levers:

- In general it is better to use levers that influence the environment or the way we in general interact with it, rather than case-by-case specific interventions; this is more beneficial to all parties and limits the risk of unintended consequences.

- Levers must be applied within the context of defined sector and capability strategies.

- Impact must be assessed thoroughly before introduction.

- Specific interventions must be targeted, contained and time-bounded.

A3.63 **Government as Regulator of industry** – on occasion, governments may need to exploit levers more concerned with controlling or restraining parts of industry to ensure access to, or control over, key IPR, capability and capacity. Equally, governments can consider where regulations can be relaxed or removed, to increase industry's profitability and agility.

A3.64 Levers that affect the way governments execute their core business are the most attractive because they impact directly on defence effectiveness and relate to core processes and specific lines of activity or programmes. They also fundamentally determine the attractiveness of the domestic defence business to industry.

A3.65 Secondary interventions can be aimed at further enabling a government's ability to:

- stimulate the overall health of the defence industry by reducing barriers to entry, encouraging participation, stimulating industry investment and stimulating technology transfer from international defence sources of adjacent industrial sectors;

- guarantee access to and control over critical technologies and capabilities where these underpin critical military capability and operational sovereignty.

A3.66 Each of these categories has within it a range of specific actions that can be taken to produce an effect. For example 'R&T support' covers a range of intervention options, from working with commercial suppliers of ICT to influence their future products ranges to direct funding of research or acquisition of specific IPR.

A3.67 Some of the levers are intended to influence the environment and can be applied at a general level. Others are intended to achieve a very particular effect and must be highly targeted.

A3.68 Different governments choose to deploy different sets of these levers; for instance, in some European countries the State continues to take significant stakes in their domestic defence industry, and control executive appointments. In the next chapter, we explore in more detail how the UK Government deploys these levers to differentiate the UK market and make it an attractive place for investors and workers to build defence companies.

Introduction

A4.1   Chapter A3 described the general shape of the evolving global defence market, and introduced the general framework of levers available to Government to influence it. This chapter illustrates some of the ways in which the Government, within this framework, seeks to continue to make the UK a attractive place to establish, grow and invest in high-technology defence businesses.

A4.2   Some of the 'levers' are in fact natural consequences of broader Government activity, for instance stewardship of the economy. Indeed, indirect levers are generally to be preferred, as they apply to all businesses and run the least risk of distorting the allocation of resources. However, while defence is becoming more international, it is far from a normally functioning market, possessing a number of characteristics which, taken together, differentiate it from other areas of manufacturing:

- national security considerations;
- few suppliers, very few legal customers;
- high levels of technological and scientific intensity;
- the central role of Governments as sponsor, principal customer and market gatekeeper;
- to different extents in different countries, legal and political restrictions on company ownership.

A4.3   All of these can affect how investors view the sector. For instance, investors are likely to require a premium where they believe Governments will intervene to secure political, rather than commercial, priorities. On the other hand, the significance of the home customer means that, if investors have confidence in the consistency and credibility of a government's approach to the industry, the market may be less volatile.

Laser technology research © BAE Systems.

A4.4   These characteristics have generally resulted in a split between a series of 'homeland' domestic markets with some international interaction, and 'export' markets for those countries without a significant defence industrial base of their own. Some sections of the supply base do however operate on a genuinely international basis. In addition, major defence programmes frequently have characteristics similar to the civil aerospace and other technology-intensive sectors:

- high cost and high risk projects;
- high value, low volume products;
- international collaboration in design and development;
- high barriers to entry;
- issues around safety critically, long-service lives and hence obsolescence.

A4.5   The market is also in general characterised by a significant element of advocacy from firms and other governments attempting to influence high value or high prestige equipment procurement decisions.

A4.6   Around 165,000 people are directly employed in defence manufacturing in the UK, with a further 135,000 people employed indirectly in supply chain activity[1]. DTI estimates that the average labour productivity is £55,000 value added per employee, contributing nearly 0.5% to GDP. It is a technologically intensive business, and an important part of the manufacturing sector which is a priority area for Government. There is widespread acceptance that the UK cannot compete on low wage activity, nor should it seek to do so. The future of UK manufacturing therefore depends on continually moving up the value chain by raising investment, R&D, skills and productivity, and Defence and the UK's manufacturing interests are closely aligned in this respect.

Government as an investor

A4.7   Investment in defence-related technology is critical to retaining access to cutting-edge military capability, and the next chapter discusses in detail the MOD's research & technology activities. This however fits within a broader framework of Government support to science, technology and innovation. The Government's target is to raise the overall level of research and development (R&D) investment in the UK from its current level of 1.9% of GDP to 2.5% by 2014, and our approach to this is described in the Government's ten year **Science and Innovation Investment Framework** published in July 2004, and the work coordinated by the business-led **Technology Strategy Board** (TSB) which aims to support the pull-through of ideas emerging from the UK's world class science and engineering base. The first annual TSB report[2] explains the vision behind the Technology Strategy: for the UK to be seen as a global leader in innovation and a magnet for technology-intensive companies. This also takes into

---

[1] DASA UK Defence Statistics 2005. In this context, 'direct' employment is that generated in those companies providing the product or service directly to MOD, or that within the exporter. 'Indirect' employment is that provided through the supply chain by sub-contractors or suppliers to the 'direct' contractor. The figures reflect average full time equivalent employment in year. The figures exclude MOD service and civilian personnel, and are rounded to the nearest five thousand.

[2] 'A call to action' – the Annual Report of the Technology Strategy Board 24 November 2005, available from www.dti.gov.uk

account existing national strategies, for example, the aerospace research priorities identified by the National Aerospace Technology Srategy taking forward the work of the Aerospace Innovation and Growth Team[3].

A4.8   Some £370 million will be available to business between 2005 and 2008 as research grants, via the DTI's Research Calls, to undertake collaborative research in technologies important to the growth of the UK economy. The Government also invests more broadly in the UK science, engineering and technology base, from which the defence industry benefits; for instance, between 1997 and 2007 the science budget will have more than doubled, rising to £3.4 billion.

A4.9   The MOD is also working closely with DTI and other government departments to develop **Innovation Platforms** where central Government, Research Councils, Regional Development Agencies (RDAs) and Devolved Administrations, business and the science base can work together to use a range of technologies and policy levers to deliver innovative products and services for which there is both a strong policy need and a potentially large global market. Network Security has been defined as one of the pilots for this.

Hand assembly soldering © Raytheon.

Government as supporter

A4.10   The Government has a broad range of levers in this area, which can significantly contribute to making the UK attractive to investors in defence and more generally. Some affect long-term characteristics of the UK economy, and government can only attempt over time to influence these. However, in general, Government assists in producing an attractive environment through:

- maintaining a **stable macro-economic and political environment**;

- supporting the education and science base, as mentioned above, to maintain a highly-skilled workforce, in conjunction with industry's own investment;

- supporting an education and business environment which also produces strong **support** industries for defence, particularly in finance, business services, design and marketing;

- keeping down the costs of setting up and running businesses – e.g. at 30% the UK has one of the lowest main corporate tax rates of the world's major economies, and the cost of setting up a business is low and the process straightforward;

- **promoting inward and regional investment**: as well as marketing the attractions of the UK as a place to invest, grants may be available from the RDAs in England and the devolved administrations for Wales, Scotland and Northern Ireland to cover a proportion of eligible costs for businesses planning to start up operations, and UK Trade & Investment, a joint DTI-FCO organisation, handle applications in the first instance. Regional support is also provided by way of Selective Finance for Investment in England, which allows RDAs or DTI to consider paying some of the capital investment costs for companies in an EU assisted area[4]. In many cases, similar schemes are offered by the appropriate organisations in Wales, Scotland and Northern Ireland;

- providing **active support to exports**, including through the Export Credits Guarantee Department and, specifically for the defence industry, the Defence Export Services Organisation, as well as political support to specific export campaigns;

- encouraging innovation through **fiscal incentives**, e.g. the R&D tax credit[5], a company tax relief that can either reduce a business' tax bill, or for some SMEs, provide a cash sum;

- the **Manufacturing Strategy**, updated in 2004, and the Manufacturing Forum – jointly chaired by the Minister for Industry and a senior industrialist, with a two-year remit to deliver practical actions to improve manufacturing in the UK. The Government provides this support in recognition that high value exports benefit the UK economy. Manufacturing exports, to which the defence industry is a major contributor, account for two thirds of all exports;

- support to defence-applicable skills. For instance, the Education Secretary and Trade & Industry Secretary announced on 31 October 2005 a **Manufacturing Skills Academy**, to open by September 2006, creating a single point of access to deliver globally competitive skills for UK manufacturing. In addition, the Government is working with Science, Engineering and Manufacturing Technologies Alliance (SEMTA), a sector skills council, which has launched a **Sector Skills Agreement for Aerospace**, together with a costed action plan that will be funded from both public and private sources. A similar agreement for Marine industries is expected in early 2006;

- providing the **Manufacturing Advisory Service**, aimed primarily but not exclusively at SMEs, to provide diagnostics and advice through regional centres of manufacturing excellence[6]. The Service has helped companies generate an additional £188 million in value added since 2002;

- more general support to business; e.g. the DTI runs a number of **business support schemes** that may be suitable for defence companies. It also offers flexible products from its business support portfolio aimed at SMEs, including the Small Firms Loan Guarantee Scheme, Support to Implement Best Business Practice, Grants for Research and Development, and Grants for Investigating an Innovative Idea (reimbursing consultancy costs for advice on how to exploit an innovative idea).

---

[3] Aerospace Innovation and Growth Team's (AeIGT) 'National Aerospace Technology Strategy: Implementation Report' followed the AeIGT's 'An Independent Report on the Future of the UK Aerospace Industry: Executive Summary' published June 2003 (www.aeigt.co.uk). The 'National Aerospace Technology Strategy: Implementation Reports' followed the AeIGT's report.

[4] The level of grant available is determined on a case-by-case basis, linked to the quantity and quality of jobs created or sustained.
[5] Introduced for SMEs in 2000 and extended to large firms in 2002; SMEs have to date received £778 million and the large company credit is worth some £400 million each year.
[6] Funding is split equally between the DTI and RDAs, and £34 million has been committed in total between 2005 and 2008.

A4.11 Government could also act as a supporter in situations where, for example, the adjustment costs from defence restructuring fell disproportionately upon a particular region of the country. However, that assistance would be best addressed through regional policy and targeted assistance, such as retraining programmes, rather than through the distortion of defence procurement decisions.

Weapon interfacing electronics © Ultra.

A4.12 We noted in this list encouragement of inward investment. We recognise the very positive contribution made by some firms in developing the UK defence industrial base, and although other nations do not operate as open a market (for either investment or procurement) as we do, we do not believe that further protection for the UK market is the answer. However, we will continue to work with individual companies where they identify that they are being prevented from making the acquisitions they want overseas, and more generally seek to open up other markets which remain heavily protected.

Government as regulator

A4.13 Government, as a regulator and de-regulator, can have very significant direct impact on the business environment – and the potential distorting effects of changes to regulation need to be carefully assessed. However, there are specific controls in place.

A4.14 In particular, the Government has a range of instruments to regulate acquisitions and mergers to ensure fair markets and security of supply. Merger policy in the UK is regulated under the Enterprise Act 2002, which came into force on 1 May 2004, and the EC Mergers Directive 04/139. The Enterprise Act effectively took politics out of merger decisions, leaving the Office of Fair Trading ('OFT') and the Competition Commission to take decisions as the independent competition authorities[7]. The Act allows the OFT to refer any merger to the Competition Commission if it believes that a lessening of competition would result in any given market within the UK (or any part of the UK), require undertakings without reference or accept the merger without reference.

A4.15 The Enterprise Act also allows the Secretary of State for Trade and Industry to intervene on public interest grounds, including national

[7] *The OFT will investigate competition issues of mergers where the annual turnover of the enterprise being merged/acquired is greater than £70M, or if the merger/acquisition will result in a share of any given market or markets of 25% or more. Mergers involving two or more member states in the EU, and with turnover in excess of 5 billion Euros, come under the jurisdiction of the European Commission under the EC Mergers Directive. However, the OFT retains the role as the competent authority in the UK by liaising with the EU Commission on mergers that fall under its jurisdiction.*

security. In such cases, Government often seeks **undertakings** from acquiring companies on retaining defence capabilities in the UK. In addition the Government holds special shares in BAE Systems and Rolls-Royce in order to protect some vital defence industrial capabilities.

A4.16 Other aspects of regulation include the setting and enforcement of strategic export controls, discussed in chapter A6; the MOD's policy towards Intellectual Property; and requirements for the vetting or nationality of individuals within business before they can receive sensitive information. In some cases, this applies at the company level – cf. List X, as discussed in the Counter Terrorism Chapter. It also includes the Government's approach to the labour market, where the UK's regulations are the most flexible in Europe.

Government as a customer

A4.17 How MOD behaves as a customer is most probably the critical factor in making the defence industry attractive to investors and workers. Much of this document sets out how we aim to align our and industry's behaviours and processes, in order to ensure the capability requirements of the Armed Forces can be met, now and in the future. In particular, we can:

- increasingly accept COTS technology, standards, architectures, products or services, and use open standards and architectures, wherever security considerations permit, to reduce cost to MOD, limit the risk of obsolescence, extend the market for industry, and open it to the widest range of suppliers possible;

- consistently act with a view to through-life capability management and take into account the need to sustain sovereign capabilities;

- select acquisition models depending on the specific circumstances, to deliver the best long-term value for money. This includes promoting a sustainable industry that secures onshore the industrial capabilities we require to maintain appropriate sovereignty.

A4.18 The continuing changes needed to deliver this approach are summarised in Part C. However, it is also important to recognise that in some areas the relevant industry is broader than Defence, and includes for instance the 'blue-light' services and other Departments and Agencies. The Government is keen to promote the spread of best practice across Government and the development of coherent approaches to procurement, and DIS is one part of this. For example, we recognise in this strategy the need to work more closely with other Departments and Agencies to produce sustainable high-grade cryptographic industrial capabilities in the UK, including by considering the amalgamated demand across Government for cryptographic products and services.

Government as planner

A4.19 As chapter A3 described, we are dependent on suppliers with very different characteristics for the supply of defence products and services. Some serve us alongside a range of wider civil and overseas markets; others are highly dependent on the domestic defence market. Our actions as planners can however help encourage these suppliers to remain effective and efficient, with most improvement likely where companies both have the resources to engage in strategic planning, and have significant potential or actual business with the MOD relative to other customers. Nevertheless, focusing solely on such suppliers would risk discouraging new entrants or limiting competition.

A4.20 We explain in more detail in Chapter A8 how we intend to **make as much information available to industry as possible**, ranging from publishing information on equipment projects annually, to deeper joint planning activity where this is suitable and can realise substantial benefits on both sides without excluding other companies inappropriately. We are also, with this strategy, publishing a guide to make clearer for industry who does what within the MOD and who therefore is empowered and **able to speak authoritatively** to industry on particular topics. In sum, we plan to move substantially to a more open and consistent face towards industry. In return, we expect industry to treat such discussions in a mature way, respecting our confidences and committing to a **mutual** sharing of information where this is appropriate.

A4.21   In addition, in this strategy we identify the industrial capabilities we wish to retain onshore for national security reasons. Where these may be at risk, we will work together with industry to investigate sustainment strategies. Some other nations are taking similar approaches. We believe that this strategy will nevertheless be distinctive, by being endorsed across Government, and set against a hard-headed assessment of the current industrial reality, recognising that in some cases while we would like to maintain an onshore capability, it will not be possible to do so unless industry works with us to address serious issues about its current sustainability and productivity.

Conclusions

A4.22   This and the preceding chapter have discussed the global and domestic defence industrial landscape, and how the Government supports a business environment that makes the UK an attractive place for the defence industry. In sum:

- **a greater proportion of our overall business is available to industry** than in any other major defence nation, and growing expertise in the combination of systems engineering expertise, agility and supply chain management required to deliver **through-life capability management** gives the UK defence industry a comparative advantage;

- a **sophisticated demand** for high-value products which have to stand up to active service, and which then, having been proven in this way, are easier to market to export customers;

- an **open market** and **diversity of suppliers** which encourages innovation, new entrants (including in information and communications technologies) and inward investment;

- we are **open to new procurement models**, including long-term partnerships, which incentivise industry to drive down costs but allow increased profits for good performance and delivery;

- in addition, as a customer, supporter, regulator, investor and planner, the Government helps sustain an attractive **overall business environment**, including:

- **a stable macro-economic and political environment**;

- leadership in **science & technology** and **manufacturing**;

- **low costs** (including low corporate tax rates, and business-friendly regulation including flexible labour laws);

- strong **support industries** in finance, business services, design and marketing;

- a highly skilled and flexible **labour force**;

- a **transparent** business environment that encourages **fair competition**;

- **specific support** to the Defence industry, including the Defence Export Services Organisation (DESO);

- increasing **sharing of information** to allow industry to plan with confidence and to attract investment;

- through this DIS, a **clear articulation of those capabilities we wish to retain onshore**, alongside clear assessments of what needs to change if these are to be sustainable, including through further **joint work with industry**.

A4.23   The Government believes this amounts to a unique, distinctive and attractive environment for the defence industry. If the opportunities are grasped and the challenges tackled, we believe the UK defence industry (which will continue to include foreign-owned companies) offers the chance of long-term prosperity, focused on delivering the high-value products the Armed Forces need.

In order to meet the challenges of the future we must be able to derive the full benefits of advancing technology'[1]

New Generation Service Respirator.

Introduction

A5.1   Defence is underpinned by increasingly sophisticated technologies[2] and the UK's battle winning military capability remains heavily dependent on the development, exploitation and insertion of world-class technology. UK Government, industry and university research and technology (R&T) effort remains critical at a time of uncertain threats, as illustrated by the diverse tasks demanded of our Armed Forces since the end of the Cold War. Although the nature of conflict remains dirty, dangerous and deeply personal often with no substitute for 'boots on the ground', the UK needs to stay ahead in technology against both conventional and novel threats, such that we can quickly develop counter measures and solutions as new threats emerge. A strong and innovative science and engineering base in UK Government research agencies, industry and universities is essential to meet this need. A fundamental task is the identification of the key technologies in which the UK should retain an international lead, due to their significance both to sovereignty and to the competitiveness of our defence industrial base. Tackling this task is made easier if we have a clear understanding of our current defence technology strengths as compared to international standards.

R&T investment is critical to the delivery of battle winning capability

A5.2   Well targeted investment in R&T is a critical enabler of our national defence capability; it strengthens innovation in our defence industry, produces more capable equipment for our forces and underpins our ability to operate with high technology allies like the US or France and lead an ad hoc coalition (e.g. of European nations).

A5.3   Technology is a key driver for change in the modern world and is crucial to network enabled, adaptable and rapidly deployable forces. Technology is one of the drivers of productivity and underpins much of the UK's productivity and success in the past few decades. This is recognised by the creation of a 10 year R&T UK Government strategy[3] to raise public and private sector investment in R&T. The results of Government R&T are often transferred into industry or carried out under contract by industry or universities. It aims to drive innovation in the UK, thereby generating greater capability for our forces. Improving equipment quality will also have a positive effect on UK defence export performance.

Benefits R&T investment has brought to military capability

- thermal imagers/night vision goggles giving night time combat advantage to our forces;

- Chobham armour, a key factor in the success of US and UK forces in both Gulf conflicts, offers the best armoured fighting vehicle protection available;

- sonar 2193, which has greatly enhanced the ability of the navy to detect traditional and stealthy mines;

- technologies to detect and counter terrorist use of explosive devices, such as the Carver remote controlled robot;

- the flight propulsion control system to make carrier landings easier for the Joint Strike Fighter;

- stealthy materials such as tiles for acoustic stealth on our nuclear submarines;

- better respirators, detectors and improved vaccines to protect our troops from chemical and biological attack.

A5.4   The MOD and its partners must continue to meet the needs of the Armed Forces and should therefore focus attention on:

- technology that can and should be inserted into future capabilities, directly improving the delivery of military effect;

- technology that will enhance the delivery of capability and decision making;

- scientific/technological advances in which the UK Defence needs to sustain a suitable level of capability in order to act as an intelligent customer;

- technology judged to be of emerging relevance to defence;

- R&T capability to inform the identification and analysis of threats.

---

[1] *'Delivering Security in a Changing World - Defence White Paper 2003. Chapter 3.4*

[2] *Delivering Security in a Changing World - Defence White Paper 2003*

[3] *Science &Innovation Investment Framework 2004-2014 - HM Treasury*

A5.5   It is important to ensure that we have skilled R&T staff embedded in and supporting the acquisition process from start to finish, making the best decisions and choices with the resources available, and therefore ensuring the effective delivery of battle winning capability.

Defence R&T is vital to:

- meet defence challenges;
- deliver cost-effective military capability;
- counter new or emerging threats;
- enable effective acquisition processes;
- support national competitiveness.

Challenges facing R&T in enabling military effect

A5.6   Current and future threats may increasingly involve the use of asymmetric tactics, requiring the MOD and industry to retain flexibility and the required technological advantage to overcome these threats. This is compounded by:

- the proliferation of technologically advanced systems, readily available on the open market that can greatly enhance the military effect that can be achieved;

- chemical, biological, radiological and nuclear materials and expertise proliferation;

- the extremely rapid pace of technological change.

Multifunction Electronically Scanned Adaptive Radar (MESAR) test facility - QinetiQ.

A5.7   The international environment in which technology based (defence and civil) industries compete is evolving rapidly. Many nations with growing economic wealth, such as China and India, are now investing heavily in R&T. Although UK investment in R&T has risen in cash terms, it fell as a proportion of GDP from 2.3% of GDP in 1981 to 1.9% now. There exists a risk that in the coming decades the UK could fall behind both our key allies and emerging economies in our ability to support sophisticated and competitive technology based industries. We could become increasingly dependent on defence technology solutions generated by other countries, including those developed from civil applications. This is of less concern when these are our allies, but a growing reliance on products such as electronic components from other regions of the world could be a source for concern in the future. Nevertheless, the UK remains a world leader in some areas; UK's aerospace and pharmaceutical industries in particular are among the strongest in the world, with aerospace accounting for 15% of all business enterprise research and development in 2004[4].

A5.8   A recent MOD-sponsored study analysing 11 major defence capable nations has uncovered a highly significant correlation between equipment capability and R&T[5] investment in the last 5-30 years as shown in Figure A5(i). It shows that there is a simple 'you get what you pay for' relationship between R&T spend and equipment quality, with a sharp law of diminishing returns, and that R&T investment buys a time advantage over open market equipment. The UK is currently in a relatively good position, reflecting a high R&T expenditure in the past but the gap with the USA is growing, which may affect UK/US interoperability. China is expected to grow rapidly to an estimated R&T expenditure level equivalent to the UK by around 2020.

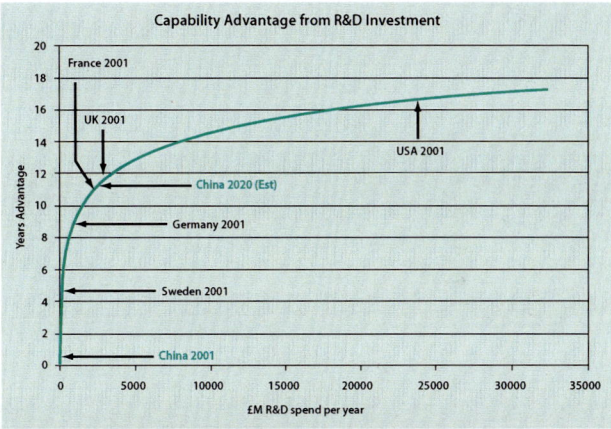

Figure A5(i) – Capability Advantage from R&D Investment.

A5.9   In order to maintain appropriate equipment quality (and hence military capability) sustained and targeted investment in MOD and industry R&T funding remains important. Focused R&T also provides battle winning capability by supporting other important activities across the MOD. These include the analysis of intelligence, current and future threat analysis, operational analysis of future force structure plans, policy formulation and equipment acquisition decisions.

A5.10   Giving our Armed Forces their critical edge has always been heavily dependent on the development, exploitation and insertion of world-class technology. This has traditionally been through MOD funded and conducted R&T, but recent years have seen the migration of part of this work to the

[4] *UK Business Enterprise R&D (BERD) is R&D performed in the UK within business enterprises, whether funded by industry themselves, from overseas, or by Government. Defence BERD rose by 13% in 2004, according to Office of National Statistics figures*
[5] *Research and Development (R&D) as defined by the Frascati Definitions of the Organisation for Economic Cooperation and Development (OECD).*

private sector. The future will demand a balance of continuing MOD and industry R&T in key and emerging areas. Military off the shelf (MOTS) and, where appropriate, commercially off the shelf (COTS) solutions can contribute to meeting capability challenges, but cannot be the complete answer.

A5.11   There will be an enduring need for MOD R&T expertise to support all aspects of defence capability and equipment acquisition, ownership and deployment. We need to rapidly prototype technologies for insertion into capability and understand how logistics and equipment support can be made more efficient and effective through the use of technology. Through all of this, the increased use of experimentation and demonstration should enable us to determine how best to meet these challenges. To achieve this we need to assess technology, conduct analysis and integration, improve the acceptance process and through life support process, including obsolescence management.

A5.12   The in-service life expectancy of major platforms is increasing. New equipment and systems need to be even more flexible to meet unpredictable demands, adaptable to ensure connectivity in a network enabled world, and capable of regular upgrade. We must find ways to exploit the global market and insert technology into military capability to keep up with the high rate of refresh within many sectors.

A5.13   MOD, in collaboration with the UK Council for e-Business (UKCeB) is undertaking a number of R&T projects aimed at establishing shared information environments with industry. Recognising that many of the major platform projects are multinational, the Transatlantic Secure Collaboration Programme (TSCP) is focussed on providing a framework for electronic collaboration that meets national and business constraints on information sharing. Initially providing a secure mail capability, the programmes goal is to provide a secure federated computing environment that can be applied to a wide range of projects.

A5.14   There is also a need for a better understanding of the human factors that determine information assimilation and subsequent action, and the need to design information networks across different components of military capability.

The way ahead

Responding to the challenges

A5.15   We have seen that Defence R&T investment is a critical enabler for military capability and the competitiveness of our defence industry. It also impacts on defence sales and the UK economy via the spin out of civil technology companies arising from defence R&T spend[31]. Employment and training opportunities in high-technology also provides broad support to wealth creation in the UK. We have recognised that the environment we operate in has changed. Industrialised nations remain very strong in the field of R&T, but developing nations are investing significantly, with civil sector research leading in some dual use areas, specifically ICT. In addition, the nature of the threat has changed and we can expect opponents to use MOTS and COTS systems (often combined) to challenge us and our interests. In this new environment, we will have to horizon-scan effectively technology advances (both threats and opportunities) and to access and exploit the best technology to give the UK Armed Forces the military capability required. In order to move forward, there are several areas that we need to address, including how we maintain our capability, how we access technology for exploitation to meet our requirements, and how to improve the way in which we conduct acquisition and exploitation.

Setting and aligning national defence R&T priorities

A5.16   In Part B of this strategy an analysis of the cross-cutting technical priorities for defence capability identifies those areas in which the UK needs to sustain or develop technological strength; there are areas of technologies with emerging defence relevance, which we need to watch. And we also need to look at where the convergence of key technology fields like ICT, life sciences and nanotechnology might create new threats or capabilities with defence relevance.

A5.17   The majority of MOD's research programme is broadly aligned with the Department's needs. However, we must ensure that our future research programme remains strongly aligned and is able to address rapidly the changing defence and security situation. We will ensure that our Defence Technology Strategy (DTS) continues to be aligned with defence policy and capability requirements. We will work with industry and universities to update the DTS and also engage in joint horizon-scanning activities to identify those emerging technologies with potential defence relevance.

Key R&T challenges for UK Defence:

- maintain technological advantage to counter emerging threats;

- sustain investment levels to maintain our relative global position;

- develop knowledge management and systems integration skills in the defence sector so that technologies can be matured and integrated into war winning systems for the future;

- recruit/retain skilled people to act as the MOD intelligent customer for R&T acquisition to meet defence needs;

- develop design and acquisition processes to enable technology insertion through equipment life;

Research and Development at Dstl.

---

[31] *The extent and value of spillovers between the defence and civil industries is subject to much to debate and there are many anecdotal examples of spillovers either way, but the overall picture is not clear. MOD, DTI and HMT will be conducting further research in this area.*

Maintaining the quality and competitiveness of national R&T

A5.18　Having agreed the priorities it is important that UK Government and industry work together to invest and maintain those key areas and the skilled R&T people we need to work in them so as to maximise the value from our investment.

> **Working more effectively with industry:**
>
> - the MOD Technology Strategy identifies those technologies we believe are critical to defence;
>
> - through 'Suppliers Days' we present the R&T challenges we face and the effects we seek;
>
> - increasing the amount of the research programme open to competition should broaden and deepen the supplier base;
>
> - partnering with industry and universities in our 6 Towers of Excellence to shares benefits and costs and increase the pull through of technology;
>
> - Defence Technology Centres (DTCs) foster collaboration with industry and universities;
>
>   - jointly funded by MOD and industry;
>   - MOD has earmarked £90 million to the DTCs over a 5 year period;
>   - they are diverse and cover; Data and Information Fusion; Human Factors Integration; Electromagnetic Remote Sensing, and; Systems Engineering for Autonomous and Systems;
>
> - spinning in COTS technology where it meets defence needs.

A5.19　The UK Government has set a target to raise national investment in R&T to 2.5% of GDP by 2014[6]. To support this national policy, the defence sector needs to maintain investment in R&T to a level that allows us to retain world-class capability in critical areas of national importance, an attractive partner for collaboration and an intelligent customer for systems or technologies, which will include MOTS and COTS. We will increase the value from our joint investment by:

- focusing on the capabilities needed to meet national defence and security priorities.

- working in partnership to share the costs of developing new ideas and to de-risk capabilities and systems.

- understanding and exploiting value in the supply chain.
- improving the acquisition process.

Maintaining the UK skills base

A5.20　The UK Government recognises the need for a stronger supply of skilled R&T staff and the defence sector is no exception. UK industry continues to lament the shortage of skilled engineers (41% shortfall), skilled

technicians (32%) and managerial and professional skills (28%)[7]. According to the Organisation for Economic Cooperation and Development (OECD), the UK has 5.5 researchers per 1,000 people employed (below the OECD average of 6.5) compared to 7.1 in France, 8.6 in the US and 15.8 in Finland[8].

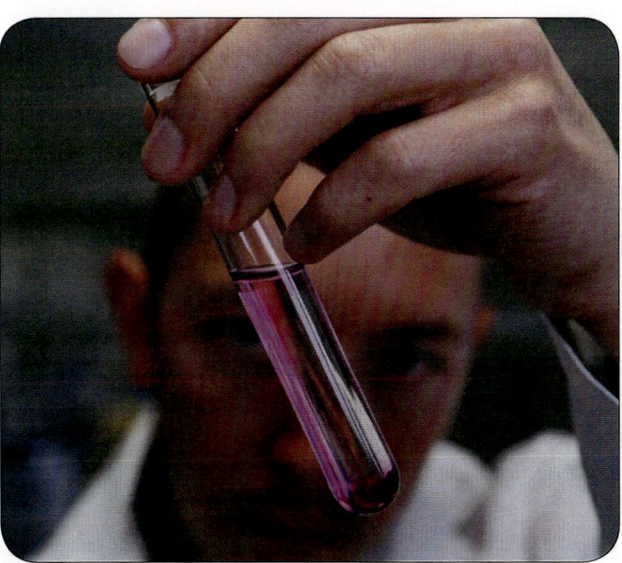

Laboratory sample testing - Dstl.

A5.21　The Roberts review[9] found that fewer students were choosing to study science and engineering disciplines and it concluded that attractive alternative careers for science graduates would constrain their supply to R&T employers and reduce innovation. The Government Chief Scientific Adviser has recently pointed out that the proportion of students studying for degrees in science has increased from 38% to 41% between 1997/98 and 2003/04, though the increases were mainly in biological and computer sciences. There has been a decrease in A-level entries in mathematics, physics, chemistry, computer science and biology of 7.5% from 1997 to 2004[10]. However, there are still relatively few students entering mathematics, hard science and engineering degrees[11].

A5.22　In MOD, our Head of Profession is charged with meeting MOD's needs for R&T staff. In order to make use of both national and international R&T, MOD and industry need in-house R&T staff with knowledge and systems integrations skills who can pull technologies together to develop capability solutions. We will also want teams which have true technical depth and world class research expertise in those priority areas we wish to lead as a nation. Increasingly interdisciplinary teams make the greatest contribution to knowledge advances.

Widening the R&T supplier base

A5.23　The MOD is moving away from doing most of its research in-house and is encouraging competition from industry and the university sector. DERA has been split into the Defence science and technology laboratory (Dstl), which focuses on core defence research that must be done in Government, and QinetiQ. In 2002/03 around 90% of our applied and corporate research was done in Dstl or QinetiQ. By 2009/10 we plan to compete around 60% of the research budget[12] that equated to our applied

---

[7] *EEF South Employer Survey 2003*

[8] *Strategic Science provision in English Universities. HOC Select Committee on Science and technology inquiry 2005.*

[9] *Sir Gareth Roberts, SET for success – the supply of people with science, technology and mathematics skills, April 2002.*

[10] *UK must go on promoting and funding science - Nature Volume 483, 3 November*

[11] *Science and innovation investment framework 2004-2014 July 2004*

[6] *Science and innovation investment framework 2004-2014 July 2004*

and corporate research programme, retaining only 35% in Dstl, as depicted in Figure A5(ii). QinetiQ is free to compete with or partner with companies in bidding for competed research, and has won work that MOD has competed.

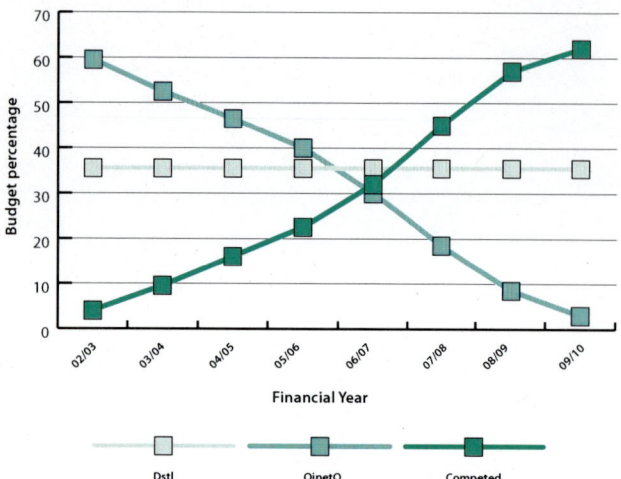

Figure A5(ii) – Increase in competition in the MOD Research Programme.

A5.24   In addition to developing the supplier base through competition, initiatives like Towers of Excellence and Defence Technology Centres have helped to increase partnering arrangements, share costs and improved bid quality. We are also developing the work we do through collaborative ventures: for instance we are pursuing the setting up of joint government/ industry partnerships such as an International Technology Alliance (ITA) in the realm of network and information sciences (UK/USA) and an Innovation & Technology Partnership for Guided Weapons technology (UK/France). Where elements of the R&T programme are contracted out, it remains MOD policy that, to the extent allowed by our international obligations, this work should normally be carried out onshore. This is to ensure that the UK retains and develops those capabilities required for its national defence and security. However, we shall seek to increase competition in Europe to its maximum useful extent via the use of research collaboration between governments; competition will be the preferred method for letting such contracts.

Changing the way we access R&T to support military capability

A5.25   We will work with the national and global technology supply chain to meet the demands of the military requirement. We will compete much of our research programme, which will help develop the supplier base. We will also better target our international collaboration activities to access global R&T with defence relevance.

A5.26   We will use the commercial market for our capability solutions wherever appropriate but the ability to provide an intelligent customer function for MOTS and COTS solution will persist. This transfer, and adaptation as necessary, of civil technologies, either directly or via products, into Defence is called 'Defence spin-in'. Expert in-house R&T capability will maintain the UK's credibility in areas of international collaboration.

A5.27   **Maintained fully in the UK** - This R&T capability will be of vital strategic importance to UK Defence and will require full provenance of a UK supply chain to protect our security and sovereignty.

A5.28   **Collaboration with International partners** - The role of collaboration is to support technological excellence in strategic defence areas, and to provide a wider understanding of defence applications across the technology domain. The latter is particularly important for commercially led technologies.

Technology tiers enabling Defence capability:

- maintained fully in the UK. (Full provenance and sustainability of supply);

- collaboration with International partners. (Visibility and trust in supply chain coupled with influence and knowledge gain in R&T);

- MOTS solutions. (Military systems with some provenance and sustainment of supply chain);

- COTS solutions. (No need for provenance or sustainment of supply chain).

A5.29   The MOD's international collaboration strategy will enable us to identify nations' strengths and develop partnerships that will increase UK and partner nation military capability. Around 12% of our current research programme would not be achievable without international collaboration and it is essential to sustain this if we are to continue to operate alongside high technology allies. Our main partners for collaboration are the US and European nations, specifically France and Sweden. Italy may also increase in importance, in line with the inward investment of Italian industry in the UK.

International R&T Collaboration:

- we will pursue international research collaboration where it adds joint long-term value to Defence; this may also provide benefit to industry;

- research collaboration can produce a return of up to five times the value of the UK investment;

- research collaboration with the US will continue to be of significant mutual benefit;

- European collaborative research will focus on joint industrial programmes to develop defence relevant technology;

- the European Defence Agency is expected to help identify opportunities for collaborations between European nations;

A5.30   **MOTS solutions** - There will be instances in which MOD paying for the development of technology and equipment neither represents good value for money nor is essential to maintaining national capability. In those cases we shall go to the best possible source to meet our needs, subject of course to security considerations.

A5.31   **COTS solutions** - These will meet many of our technology needs from the broader industrial and university base. Furthermore as the global investment in R&T continues to increase, and as an ever larger number of countries contribute to this overall growth, it will not be

---

[12] *NAO report Management of Defence Research and Technology 10 March 2004. (note this refers to what were the applied or corporate elements of the research programme which formed the QinetiQ assurance and not the whole research programme).*

possible for the MOD to support cutting edge R&T activity across all areas of R&T relevant to defence. However, it is unrealistic to assume that the majority of future defence needs could be met with COTS solutions.

Advanced mobile communications for the battlefield - QinetiQ.

Defence 'Spin-In' and COTS:

- globally, civil markets are an important driver of science and technology. It makes sense to harness these civil technologies for defence use, an example being the military use of ICT;

- an active programme of technology watch is essential to ensure that we identify these technologies;

- most 'spin-in' is the purchase of COTS products, rather than the utilisation of civil intellectual property;

- the difficulties with integrating commercial civil products into complex military systems can often lessen the attractiveness of a COTS solution;

- our R&T programme focuses strongly on those technology areas in which the civil sector is unlikely to produce the solutions we need;

- where our research programme generates technology with dual-use we shall continue to ensure that is exploited to its fullest potential, through review of our intellectual property policy.

A5.32   Figure A5(iii) represents a complex modern military platform through a small number of top level systems. These systems will contain a balance of bespoke, MOTS and COTS equipment. Whilst COTS systems will be heavily utilised to ensure cost effectiveness, there will always be significant elements that must be bespoke defence technology led. e.g. weapons, countermeasures, sensors, combat systems, aspects of propulsion and elements of platform design. It must be noted that the balance of design and development effort can vary significantly depending on the project. The key is to understand where civil COTS can be integrated alongside defence technology to give value for money and military capability.

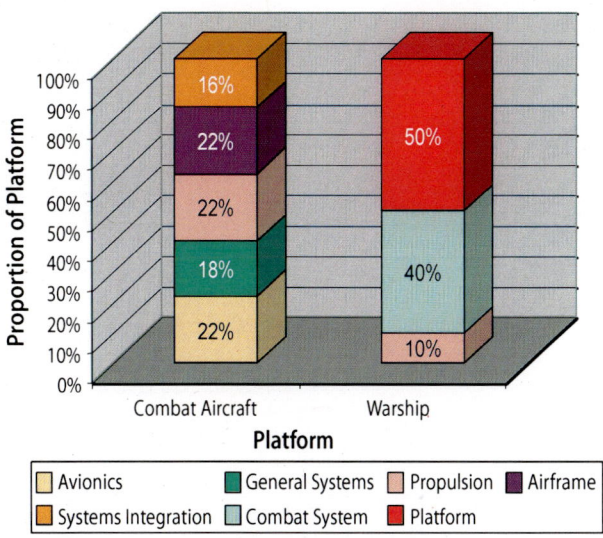

Figure A5(iii) – Breakdown of military platforms by system type (balance of design & development effort).

Ensuring R&T feeds into and improves the defence acquisition process

A5.33   R&T must also be used to improve the acquisition process by supporting our ability to make better decisions and enabling us to identify more cost-effective ways of meeting capability needs. Appropriate and targeted investment in this area can reduce risk throughout the acquisition process, through analysis to generate informed options and an increased understanding of technological risks, the use of exploitation plans and technology demonstrator programmes.

Exploitation plans and technology demonstrator programmes

A5.34   We should seek to realise the value of innovation by exploiting it in new equipment or new processes. Equipment related MOD R&T programmes should include a greater emphasis on the development of demonstrators[12]. This requires better exploitation plans from the start of a research project, which will be used as a criterion for the provision of further funding. This will help the MOD and industry determine where funding is most effectively allocated and allow better alignment of funding between the Equipment Plan and the supporting R&T programmes.

Effective technology insertion in acquisition

A5.35   The time between major platform procurements is increasing, as is the proliferation and rate of change in R&T. The UK must increase the pace of technology insertion, drawing upon advances in the defence and civil sectors. This will allow us to respond to both evolution (the norm) and revolution (the exception) in capability. We should look to open architectures that facilitate incremental technology insertion (i.e. 'plug and play'). Platforms and systems should be designed with upgrade and flexibility in mind, noting that new roles for existing equipment will be identified to respond to changing threats. We cannot keep pace with US investment, but we must ensure that we are interoperable with them. If we can find a way to plan and acquire our systems and platforms with technology insertion in mind then we can better sustain the capability of our military equipment in a more cost effective way. This will maintain the standing of our military capability where it needs to be, relative to other allies and competitors. In seeking to achieve an increase in the pace of take-up of appropriate technology, the UK will continue to draw upon advances within the defence and civil R&T sectors. We need to identify those technologies that are likely to evolve rapidly in order to target the areas we need to design for modularity and insertion.

[12] *The Management of Defence Research and Technology Part 4 NAO Report March 2004*

PCR minilab a device to rapidly identify biological agent present in sample - Dstl.

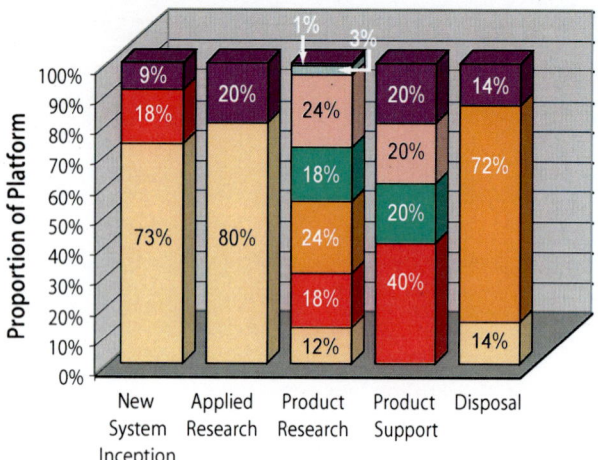

(a) Breakdown by technology phase

Value in the technology supply chain – a key power-system component showing the different types of organisation engaged at various stages in the product's lifecycle, highlighting the employment of a wide range of specialist technological providers in the development of a single component within a complex system

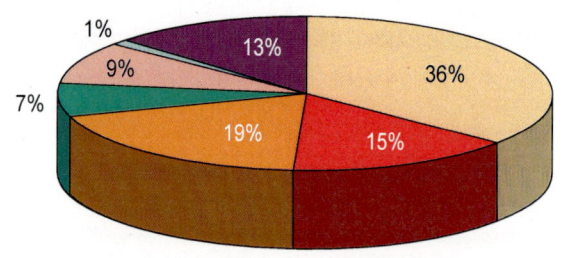

(b) Breakdown across whole-life

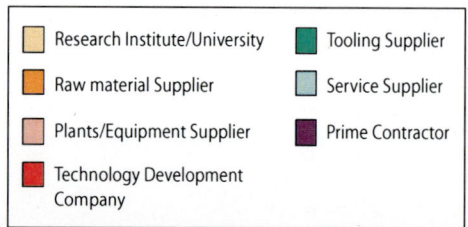

Figure A5(iv) – Organisations involved in through-life capability sustainment of a propulsion system component.

Understanding and exploiting value in the technology supply chain

A5.36   We will work with industry to gain a better understanding of the nature and structure of the technology supply chain from prime contractors through to the lower tiers. We will continue to meet most of our major technology and acquisition requirements through large prime contractors, given their expertise and experience in systems integration. However, in today's world of rapid technology change, there also needs to be a more effective engagement with the so-called 'lower-tier' suppliers; namely the small and medium sized enterprises (SMEs) and universities who are often involved in the development of very novel technologies and materials. We need to understand value at all levels, highlighting the role of innovation through the supply chain. One example of the varied technology suppliers to a major prime contractor developing a key component in a propulsion delivery system is shown in Figure A5(iv).

A5.37   The critical role of the prime/system-integrator organisation is to manage the overall design, concept and architecture of the system and the sub-system technologies. A key technological component may lie in a SME and it is important that both the prime contractors and the lower-tier technology sources are supported to ensure access to innovative technology.

A5.38   The MOD, major defence industries, universities and the DTI need to make a combined effort to identify innovative SMEs and their capabilities and improve means of engagement with them. It is important that SMEs are made aware of the opportunities available to them with defence as a possible market for their innovation and technologies.

Innovation and technology to enable through life capability sustainment

A5.39   Greater utilisation of innovation and technology will ensure that incremental capability insertion can be achieved and that system and component obsolescence can be overcome. Technology exploitation may be targeted at the reduction of through life cost and the reduction of the logistics and maintenance burden experienced with older equipment. These benefits could be delivered throughout the supply chain to enhance the operational availability of military capability.

A5.40   We will focus more R&T effort to support through life capability sustainment and continue to work with industry to develop and insert technology to increase the endurance of equipment in the expeditionary battle space and successfully deliver capability whether on land, on or under water, in the air and in cyberspace. This shift in focus, will enable through life platform capability sustainment/enhancement programmes to insert new technologies (e.g. new sensors, weapons, materials etc.) as they become sufficiently mature.

Summary

A5.41   Many of the conclusions for the R&T community and how these will be taken forward are contained in Part C and form an element of how the DIS challenges will be taken forward. Here, those challenges that are most pertinent to R&T are expanded to form an outline plan of how to meet those challenges. We must agree the R&T and capability areas we will undertake nationally and those we are happy to source on the global market from other defence industries/civil sector.

A5.42   We must align our defence priorities to meet MOD's needs and maintain the quality of UK R&T. Our R&T programme must focus on:

● R&T aligned to MOD military capability needs;

● and those emerging technologies that may be disruptive/or have defence applications.

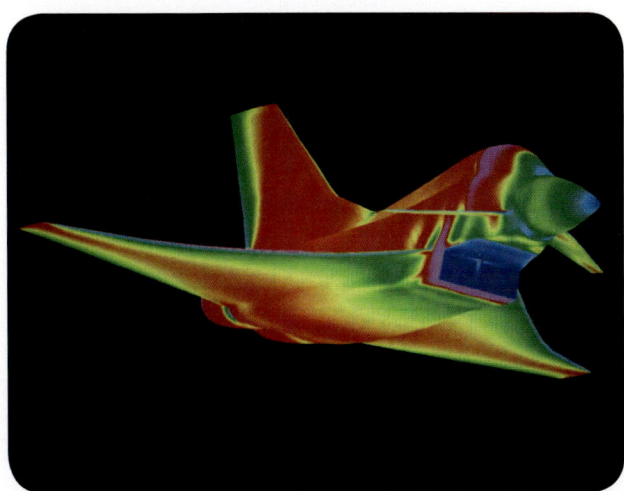

Computational fluid dynamics in Typhoon drag coefficient modelling.

A5.43   We must play our part together in supporting the Government strategy to get national R&T spend to a competitive 2.5% of GDP by 2014.

A5.44   We must work together more effectively with industry (including SMEs) and the universities, to stimulate innovation and exploit R&T to meet defence needs.

A5.45   We must grow the skills base to ensure we have the necessary supply of highly skilled engineers and scientists.

A5.46   We must develop new ways of working together to maintain national strengths, get better value from our joint investment, identify and exploit UK innovation and intelligently access and exploit the global R&T base; this will include placing greater emphasis on technology demonstration. We must consider how to improve technology insertion, thereby ensuring our systems and platforms have a battle-winning advantage.

A5.47   We need to examine how R&T might better support our acquisition process. Through the use of better decision support, adoption of new models for military effect, implementation of technology solutions it should be possible to provide more cost effective military capability and reduce technical risk in the EP.

A5.48   We must further enhance the science and technology 'literacy' and expertise of our staff. The MOD must provide career structures for scientist and engineers that recruit and retain highly qualified staff in an increasingly competitive global skills market.

Technology sharing across the Atlantic

The relationship with the USA is in general a healthy one. While the USA procures most of its equipment onshore, the UK defence industry continues to do far better than most countries in competing for US defence requirements, based on both a degree of mutual understanding and trust in our security arrangements, building on our broader strategic relationship, and mutual respect for respective industrial and technology strengths. For instance, the BAE Systems M777 155mm lightweight towed howitzer has been selected by the Marine Corps. We also welcome companies based in other countries, including in the USA, which are prepared to invest in the UK, especially where they bring with them useful knowledge and help broaden the range of potential UK suppliers, potentially helping to sustain key sovereign capabilities. In some cases, UK subsidiaries of US-based suppliers have even supplied back into the US market. That is not to say, however, that we do not continue to pursue greater reciprocity of access with the USA. In particular, and for sovereign rather than commercial reasons, we continue to strive for improvements in technology sharing arrangements required.

The size and capability of the US defence market has made it a magnet for UK-based defence industry, and an obvious source of supply of equipment to meet UK military requirements that are predicated to a large degree on US-led coalition operations. Reflecting this, the UK is currently involved in a substantial number of co-operative programmes with the USA, the biggest single programme being the Joint Strike Fighter. In addition a very large number of UK programmes, and collaborative programmes with European partners, are dependent to some degree on US technology. A number of UK companies (led by BAE Systems and Rolls-Royce, but including Cobham, Smiths, and others) have significant industrial footprints on both sides of the Atlantic.

To meet our own sovereign needs, it is important that we continue to have the autonomous capability to operate, support and where necessary adapt the equipment that we procure. Appropriate technology transfer is therefore of crucial importance. This is so for any cooperative project, but in practice difficulties have arisen particularly with the US, whose technology disclosure policy we have found less adapted to the needs of cooperative procurement than those of our partners in Europe. To reiterate, this is not about gaining competitive advantage for UK industry; it is about being confident that the equipment we buy meets the capability requirements against which it is procured and can be modified effectively to meet emerging requirements through life. We fully recognise the need to ensure that intellectual property is protected, and that appropriate measures are put in place to ensure this; security is a key issue for us, just as it is for the USA. But a certain degree of technology transfer is required if we are to be able to fully cooperate with the USA (or any other partner) on our equipment programmes. What we are striving towards is an agreed framework which facilitates this whilst ensuring that our mutual security needs are met.

The importance of transatlantic defence industrial cooperation lies in enabling both UK and US Armed Forces to acquire more effective military capabilities, at better value for money, than would otherwise be the case, and to cooperate together in pursuit of common security objectives. It is in the interest of both Governments and of both industries to improve the current situation.

Introduction

A6.1   During the Cold War, the Soviet Union supplied and dominated markets in other Warsaw Pact countries and client states for the output of its military industrial complex. There were also established defence trade links between NATO countries and allies. Following the demise of the Soviet Union and end of the Cold War, these established defence trade links broke down and during the 1990s the export market became more competitive, with for example, Russia and Poland now competing in markets alongside the USA, the UK, France and others.

A6.2   Over the last decade , the UK defence industry has won export orders worth an average of £5 billion per annum, bringing significant benefits to industry and MOD. This achievement has secured the UK as the world's second largest defence exporter after the USA. Now that the market has become more competitive, suppliers must be able to offer customers a broader range of services; the offer of equipment alone is unlikely to meet the total requirements of most customers, who now expect suppliers to offer package solutions covering equipment and support throughout the life of a product. Indeed, as we increasingly encourage our suppliers to design for through life capability management, this is an area where UK defence industry can demonstrate increasing competitiveness in the global market.

A6.3   The UK defence industry has a range of world-class technologies and products which it can offer allies. Much of this has been tested and adapted as necessary by real operations, and as a consequence potential export customers can have more confidence in its reliability and performance. UK success in overseas markets has traditionally been in major platforms, but exports of sub-systems and the provision of support services are becoming increasingly important. SMEs have an important part to play in maintaining the UK's strong market position.

A6.4.   The UK defence industry has also gained competitive advantage in export markets by offering the UK's world leading design excellence and pioneering innovation & invention skills. This is especially potent where another nation's design capability is fragile or has been lost. This type of export has the added benefit of helping to sustain our own design capability for use on national programmes.

A6.5   Defence exports bring commercial benefit to UK companies and around 20% of UK defence employment is in export work. However a UK defence industry that is able to generate significant export revenue also has **value for Defence** for a number of more specific reasons :

- Defence exports support **defence diplomacy** and in some countries may act as a key enabling activity for a bi-lateral defence relationship.

- Defence exports contribute to building local operational capability and therefore **enhance interoperability** with our own forces, especially during peacekeeping missions.

- Longer production runs also **spread fixed overhead costs**. The benefit thus accruing to industry may be shared by us in the form of lower prices on future purchases from the same supplier.

- By sustaining longer production runs and offering opportunities to develop equipment for export customers' requirements, defence exports **help to maintain key sovereign capabilities** in both production capacity and systems engineering skills, which we might otherwise have had to intervene to maintain.

Lynx.

A6.6　For these reasons[1], the Government puts considerable effort into supporting responsible defence exports, in pursuit of our broader foreign and security interests and for the direct value to Defence that they generate. This effort is led by our Defence Export Services Organisation and supported by other parts of the MOD, the Armed Forces and wider Government.

The Export Credits Guarantee Department supported £766m of defence business in 2004/2005, representing 38% of the total business supported.

A6.7　The UK is also at the forefront of promoting internationally the need to ensure defence exports are responsible. With a wide range of other partners the UK has made progress since 2003 in building support for international agreement at the 2006 UN Small Arms and Light Weapons (SALW) Programme of Action Review Conference on minimum common criteria to underpin controls on transfers of SALW (including import, export and transhipment). Although the UN Programme of Action specifically only covers SALW, and is politically binding, any principles that are agreed may eventually have wider implications than SALW. In the slightly longer term the UK is also pursuing the wider objective of a legally binding international treaty to cover the trade in all conventional weapons. The UK has set out six criteria for such a treaty, calling for it to be: - legally binding; include all conventional arms; a separate, self-standing initiative; based on core principles, which make clear when exports would be unacceptable; have an effective mechanism for enforcement and monitoring; and include a wide range of signatories, including the world's major arms exporters.

A6.8 There are also a few potential risks associated with defence exports, which need careful managing if both appropriate sovereignty and value for Defence are to be protected – for instance, of the unintended transfer of technology, and the risk that a friend now may be a foe later. To manage these risks, we have a number of tools, including the following:

● We apply Strategic Export Controls to prevent exports we believe to be inconsistent with our legal commitments and wider policy. These take into account for example the UK's international commitments, including sanctions and embargoes; respect for human rights and fundamental freedoms; the internal situation in the country of final destination; concerns that proposed exports might be used for internal repression or international aggression; risks to regional stability; national security; the recipient's attitude to the international community, including towards terrorism; the risk of diversion; the risk that the cost of the export would have negative developmental impact; plus other considerations as described in the Consolidated EU And National Arms Export Licensing Criteria. All export licence applications are considered on a case-by-case basis with the MOD, FCO and DfID advising the DTI as the licensing authority, with Cabinet-level Ministerial discussions as necessary.

● A recurrent theme throughout the DIS is that we along with the defence industry need to understand more clearly the complete supply chain needed to sustain sovereign capabilities. While exports will often sustain supply chains that would otherwise have been without business, this can be undermined by other nations' requirements for offset – i.e. that particular elements of the work on UK programmes should be subcontracted offshore or new industrial capabilities established in their territory. Other governments' directed offsets can have a particularly distorting effect on the supply chain's operation, however, and thus create sustainment issues for domestic sovereign capabilities. MOD's own approach to offset, known as industrial participation, does not specify where work should be placed, but encourages foreign bidders to use UK sub-contractors on a competitive basis.

Merlin helicopter.

New opportunities for export advantage

A6.9　The UK Armed Forces enjoy a high international reputation and there is no better recommendation when marketing UK defence systems overseas than emphasising its UK in-service credentials. It is much harder to sell equipment abroad that has not been endorsed by the UK Armed Forces.

A6.10　In general, defence industry's priority is to produce systems to meet the requirements that we specify, as the main customer. Hitherto, export opportunity for a particular product has not necessarily been factored into the earlier design stage. This means that, in many cases, UK equipment is too sophisticated and therefore too expensive for many potential export customers. This has tended to be the case for example with warships, and the UK has lost market share to competitors that have considered exportability at the early design stage of a platform. This has involved the adoption of greater modularity in design that then allows the supplier to offer a different set of options to export customers. As we move more to through life capability management ourselves, facilitated by open architectures and modular approaches, there is now an opportunity to consider at the design stage of a programme features which would enhance a product's exportability (and reduce its through life costs).

A6.11　Further, given the internationalisation of the defence industry, most products on the market now contain a mix of sub-systems sourced from different suppliers, regardless of whether they have been developed in collaboration with other nations or in response to a UK requirement. With the high levels of foreign investment now in the UK defence industrial base, there are greater opportunities for UK-based companies to work in partnership or in collaboration with overseas firms, thus giving them broader market access. For example, the recent teaming arrangement between AgustaWestland and Lockheed Martin led to the recent success of US 101 - derived from the EH 101 - in the US Presidential helicopter competition. In addition, since there remains further scope for consolidation within the global industry, there may be opportunities for UK companies to access new markets by virtue of selected merger and acquisition activity.

---

[1] *Arguments for supporting defence exports in terms of wider economic costs and benefits. e.g the balance of payments, are sometimes also advanced. A group of independent and MOD economists (M Chalmers, N Davies, K Hartley and C Wilkinson - 'The Economic Costs and Benefits of UK Defence Exports'. York University Centre for Defence Studies, 2001), examined these, by considering the implications of a 50% reduction in UK defence exports. They concluded that the "economic costs of reducing defence exports are a relatively small and largely one off...as a consequence the balance of argument about defence exports should depend mainly on non-economic considerations."*

Competition, alternative approaches to competition and presenting, measuring and delivering value for money in defence

Competition

A7.1   A sustainable and competitive UK defence industry remains essential to the delivery of military capability to the UK's Armed Forces. Open and fair competition is a fundamental component of our procurement policy to deliver affordable defence capability at better overall value for money.

A7.2   Figure A7(i) shows that over the past four years, about three quarters of our contracts, by value, have been let competitively, covering the full spectrum of procurement from major equipment projects through to clothing and supplies.

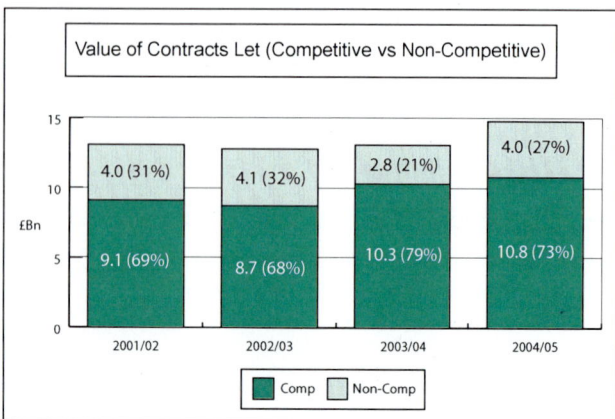

Figure A7(i).

A7.3   Procurements for non-warlike goods and services are conducted in accordance with the European Union Procurement Regulations. We will work with the European Defence Agency with regard to the procurement of warlike goods and services, to harmonise the UK's approach with the voluntary Code of Conduct on more open competition. In addition, we will continue to advertise our requirements to encourage suppliers to seek business at all levels of the supply chain. This Chapter explains how we will adopt procurement strategies to make best use of available capabilities and capacities.

A7.4   We are tackling industry's concerns about the expense and uncertainty that can arise from protracted and inefficient tendering. We are taking steps to speed up decision making and to minimise costs. For example, we will place more emphasis on targeting smaller numbers of potential suppliers with the best credentials to bid for individual requirements, rather then rely on a large panel of bidders just to maintain competitive pressure - expensive in terms of the cost of bidding, the time it takes to down-select and evaluate a large number of bids and unfulfilled supplier expectations. We will extend the use of streamlined procedures to pre-qualify potential bidders, to select early a preferred bidder and to make the maximum use of automated evaluation tools and processes.

A more flexible approach

A7.5   With increasing technological complexity, globalisation and industry consolidation, priced based competition may not automatically result in the best opportunity for successful acquisition or maintain key sovereign capabilities. This places challenges on our relationship with industry which remains firmly rooted in project performance.

A7.6   Whilst competition allows the advantage of tangible price comparison determined by market forces and the ability to compare competing proposals for compliance, it can also sometimes drive unintended behaviours and consequences for both us and industry. These may include unrealistic timescales, an over optimistic assessment of risk and hence cost, and the potential loss of flexibility for timely insertions of technology in the future.

A7.7   We will continue to use market forces where we can to determine better value for money, but defence is not a perfect market place. We will therefore adopt procurement approaches that consider the nature of the market in the relevant sector and provide the flexibility to respond to structural changes, so as to sustain key sovereign capabilities and to ensure long term value for money. The Key Supplier Management process will enable us to assess the strategic and aggregate impact of different potential acquisition choices, particularly those that have significant industrial base consequences.

A7.8   Whilst this Chapter focuses mainly on our relationship with prime suppliers, we fully recognise the important role played by Small and Medium size Enterprises (SMEs) - defined as companies employing 250 people or less - which constitute a significant core of direct suppliers to MOD. In 2004/05, just over half of the MOD contracts let were directly with SMEs, accounting for over half a billion pounds. Many SMEs play a crucial role in meeting Urgent Operational Requirements (UORs), undertake prime roles for smaller requirements, and are potentially well placed to fulfil other roles in the context of the Models described below. We are widening our supply chain focus below the prime level to identify critical sources of key capability and technology and further to encourage SME entry into a broader range of defence opportunities. Whilst the MOD already has supply chain oversight[1] provisions in place, we are developing senior level links with regional industry groups where issues of specific interest to SMEs can be bought to our notice.

Alternative approaches

A7.9   The 2002 Defence Industrial Policy (DIP) recognised that even in competitive environments there are a number of wider factors besides cost and operational effectiveness, affordability and long term value for money that will influence supplier and procurement selection. These include security of supply and the retention of key technologies and industrial capabilities, the implications for export potential, our wider policy framework and industrial participation. In addition there are procurement factors that may be assessed and these are addressed in the models set out below. These factors will be included in Invitations to Tender (ITTs) with relative weightings, and will embrace wider factors insofar as they relate to the individual procurements.

---

[1] *The MOD/Industry Commercial Policy Group Guidline No5 (Defence Acquisition, The Commercial Framework Codes of Best Practice).*

A7.10   The DIP further recognised that there are occasions when competition may not be able to deliver the best long term value for money or sustain key UK defence industrial capabilities. We will not pursue competition beyond the point where it can offer long term advantage or where the cost of running a competition is demonstrably disproportionate to the benefits that might be achieved.

A7.11   Overall, the key objective from our perspective is how to **present, measure and deliver value for money** in situations where competitions are not held.

A7.12   We will consider alternative approaches to competition in the procurement situations set out below. They are not intended to be prescriptive:

- One supplier has the capacity and capability to deliver the requirement and is chosen because it is the sole source of supply, or it is chosen on the basis of consistently high performance compared to other suppliers, or it is the only suitable supplier to sustain sovereign capabilities in industrial base or other procurement grounds.

- No single supplier has the capacity and capability to deliver the requirement and where an inclusive and willing group or groups of suppliers might be formed and sustained.

- The through life support of a capability that requires the engagement of the equipment Design Authority and/or other systems engineering capability.

- Competition exists but the procurement can readily be compared or benchmarked against similar technologies, supplies and services, or for UORs where equipment is readily available.

A7.13   The models below, and Figure A7(ii) show how we will select suppliers and undertake value for money assessments in these different procurement situations.

A7.14   The approaches set out in this Chapter are intended as a guide to our acquisition teams and industry in designing procurement strategies to achieve the best project outcomes and in making value for money and investment comparisons and decisions. The success of these approaches will depend on a number of key cultural enablers including:

- establishing and sustaining the right relationships and behaviours across the acquisition community;

- extending 'best practice' partnering approaches to appropriate MOD projects;

- engendering more openness and transparency in our dealings to secure better long term value for money for the Government, and profitability (based on good performance) for industry.

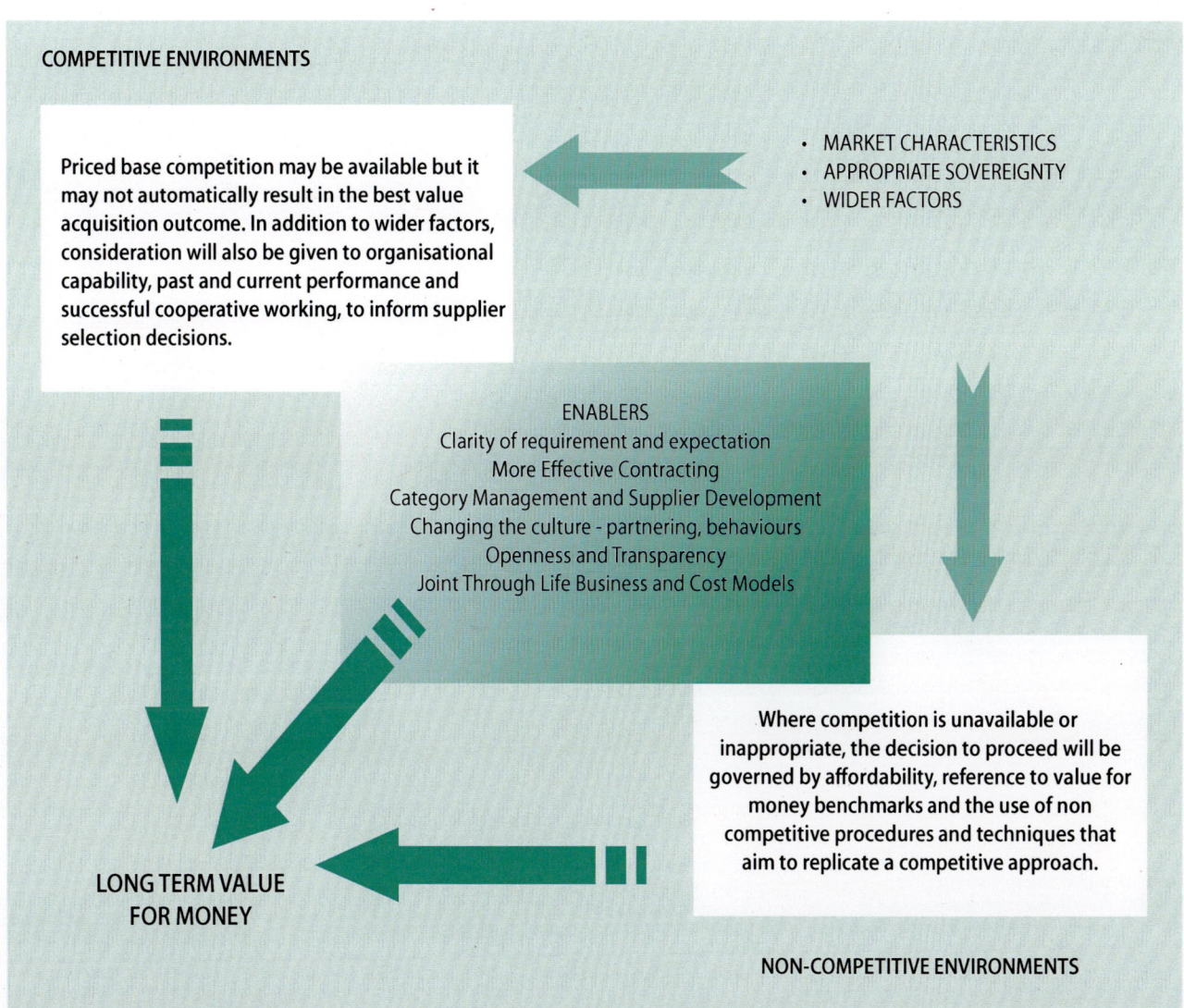

COMPETITIVE ENVIRONMENTS

Priced base competition may be available but it may not automatically result in the best value acquisition outcome. In addition to wider factors, consideration will also be given to organisational capability, past and current performance and successful cooperative working, to inform supplier selection decisions.

- MARKET CHARACTERISTICS
- APPROPRIATE SOVEREIGNTY
- WIDER FACTORS

ENABLERS
Clarity of requirement and expectation
More Effective Contracting
Category Management and Supplier Development
Changing the culture - partnering, behaviours
Openness and Transparency
Joint Through Life Business and Cost Models

Where competition is unavailable or inappropriate, the decision to proceed will be governed by affordability, reference to value for money benchmarks and the use of non competitive procedures and techniques that aim to replicate a competitive approach.

LONG TERM VALUE
FOR MONEY

NON-COMPETITIVE ENVIRONMENTS

Figure A7(ii): Factors to be considered

Staff at work at the Defence Procurement Agency at Abbey Wood.

Model 1: One supplier has the capacity and capability to deliver the requirement and is chosen because it is the sole source of supply, or it is chosen on the basis of consistently high performance compared to other suppliers, or it is the only suitable supplier to sustain sovereign capabilities in the industrial base or other procurement grounds.

Considering the approach

A7.15   For long-term projects, affordability, timescale, priorities and technology insertion plans will be communicated wherever possible so that suppliers can make informed and focused investment decisions and assess opportunities prior to commitment. Early notification will also enable suppliers to participate in the process of identifying trade offs between performance, time and through life cost.

A7.16   In devising the procurement strategy, consideration will need to be given to **factors other than price**; where there is technological uncertainty at the early stages of the capabilities lifecycle, this may not be determinable to acceptable level of accuracy. The outcome is ultimately focused on a commercial arrangement with a single legal entity.

Supplier selection

A7.17   These considerations are likely to be particularly relevant to the Assessment Phase (AP) of a project where priced based competition may not produce the best value outcome. In selecting potential suppliers, we will assess past and current performance, organisational capability, the pull through and utilisation of technology and capability, evidence of co-operative partnering and continuous improvement in productivity and reduced through life costs.

A7.18   When a project progresses from the AP to the Demonstration and Manufacture phases, the early down selection of a preferred supplier from a group of competing suppliers, may be determined by comparative reference to the factors described in the 'Assessing Value for Money' section below.

A7.19   Where there is a need to select a potential sole source supplier i.e. where the package of work itself creates or supports a critical industrial capability for reasons of sovereign control, but which is insufficient to maintain multiple suppliers, the above selection criteria will apply but with the emphasis on informing a decision about whether to proceed on value for money and affordability grounds.

Project control

A7.20   At the heart of this approach is the pursuance of improved, through life project and commercial arrangements that allow more flexibility backed up by an incremental approach to contracting, a genuine emphasis on de-risking, and the identification of key indicators to measure performance and to inform entry/exit decisions at each phase of the project.

Assessing value for money

A7.21   For sole source requirements, the ultimate assessment of value for money is taken at point of deciding whether to proceed to meet the required military capability. The value of the capability may be subject to benchmarking, but where no such benchmarks exist this decision will be against the absolute need of the capability, its relative priority and its relative affordability, throughout the capability life.

A7.22   Except for most sole suppliers, the prospect of follow on work will be a powerful incentive; the supplier's demonstrable performance and behaviours at key contract stages and the flow of sound through life value for money proposals/solutions for follow on phases will be a determinant to the placing of further packages of work. Where they exist, benchmarks derived from competitive sources will also be used to inform the selection decision.

A7.23   In the absence of competitive price pressure, it would be essential to adopt a robust approach to the assessment of cost. Parallel working by MOD and industry estimators leading to agreed estimates of cost would reflect the close and open working relationship necessary in this approach.

A7.24   It may also be sensible to use joint through life cost and business models, in addition to backwards-looking open-book accounting; in the early stages these models will inform planning and cost assumptions, providing the basis for project cost estimates. During its life the model will be actively maintained to inform subsequent business case and investment decisions and support project governance. They will also allow 'what if' analyses for changes to the project and periodic value for money assessments against external benchmarks.

A7.25   Inherent in the model construct is the joint commitment, which will be enshrined in the commercial arrangements, progressively to populate these models with the full costs of ownership as they develop and mature.

Model 2: No single supplier has the capacity and capability to deliver a requirement and where an inclusive and willing group of suppliers might be formed and sustained.

Considering the approach

A7.26   The focus falls on establishing a successful engagement with a Prime Industrial Group, with the right culture and behaviours and a willingness to work in an open, partnering environment to drive value for money throughout the life of the capability. This will include a commitment by the Group fully to share cost data and the business model through which shareholder value is to be delivered whilst respecting commercial confidentiality. The Group will seek flexible solutions with open architectures to provide for the best capability outcome, and make full use of COTS procurement wherever this gives better value for money.

A7.27   In circumstances where we seek to acquire an integrated set of military capabilities, for which we may not have the necessary domain knowledge or expertise, we may partner with a systems integrator. The systems integrator will work with us to establish the optimal mix of capabilities and suppliers to meet the requirement, to make trade off decisions and to take a wider view of network enabled issues.

Supplier selection

A7.28   In many circumstances there may only be one Prime Industrial Group which provides the breadth of delivery skills and industrial capability and capacity to meet the anticipated requirement. It may be necessary for us to act as facilitator during the period which leads to the formation of the inclusive group. The criteria for selecting eligible suppliers will largely be the same as that for Model 1.

A7.29   In forming the Group, a list of potential second tier suppliers may be drawn up. Contracts Bulletin and, where appropriate, Official Journal of the European Union (OJEU) advertising will be used to keep the supply chain advised of future competitive opportunities.

Assessing value for money

A7.30   As with Model 1 except that the procurement would be through co-operative working with the Group, in the spirit of openness and transparency, where there would be open and full discussion of all relevant data by all parties on an on going basis. This will not preclude the potential to compete individual systems or indeed elements of the whole at various stages in the programme, subject to the wider factors as discussed in chapter A9.

A7.31   Within the Prime Industrial Group we would wish to encourage commercial arrangements that share risk and rewards and incentives to drive costs down and to seek opportunities for improving value for money through the life of the project. We will give therefore renewed emphasis to Gainshare, Target Cost Incentive Fee pricing and the use of risk adjusted profit according to the type of work carried out under the contract.

Model 3. The through life support of a capability that requires the engagement of the equipment Design Authority (DA) and/or other systems engineering capability.

Considering the approach

A7.32   With the approaching relative downturn in the procurement of large new capability platforms, long term support is increasingly seen by many suppliers as the primary source of revenue in the defence market in the future. Whilst the conventional position is to seek to compete these requirements where it is practicable, and there will be instances where this still represents the best option, we will consider alternatives to competition where they offer the prospect of better value for money.

A7.33   This is particularly pertinent for legacy systems where the DA may have been appointed during the equipment procurement stage and would have acquired intellectual property, extensive knowledge of the design and expertise in its application. Thus the DA is best placed to be able to interact effectively with us to manage technical risk and trade off design investment with other aspects including reliability, safety and maintenance. In some cases the relevant systems engineering capability may be vested in a broader entity than the DA, especially when upgrade or improved support opportunities require the insertion of new technology, but it is very unlikely not to include the DA.

Supplier selection

A7.34   Generally, the preferred supplier would be the appointed DA, subject to confirmation of competence, capability, capacity, financial status and willingness to engage in an open cooperative partnering arrangement. Maintaining this relationship is of course dependent on demonstrating acceptable performance and satisfying adequately the value for money and affordability considerations.

Assessing value for money

A7.35   The challenge for us and industry is to find mutually acceptable, robust commercial arrangements that incentivise delivery of the required support at minimum and continuously decreasing cost and improved long term value for money. In so doing, we would wish to transition from arrangements that reward volume and the cost associated with that volume, to one which rewards the active management of risk and the value it brings to defence - contracting for availability and /or other aspects of military capability.

A7.36   In return for consistently high and innovative performance, industry might reasonably expect the prospect of greater reward with a better certainty of revenue. The rebalancing of risks would almost certainly extend Industry's reach and influence into support areas traditionally managed by us such as maintenance, stock control, storage and distribution. For upgrades, this might also include opportunities to work collaboratively on capability management.

A7.37   Inherent in these arrangements is a willingness by us and industry to build and sustain effective partnering relationships, to work to a clear purpose and in an open and transparent environment, to share performance and cost data and to evolve a clearly understood business model that will incorporate the full costs of ownership. These mirror the concepts detailed in Model 1.

Model 4. Competition exists but the procurement can readily be compared or benchmarked against similar technologies, supplies and services or, in the case of Urgent Operational Requirements (UORs), where equipment is readily available.

Considering the approach

A7.38   Recognising that competitions are costly in terms of time and effort to both us and industry, it may dispense with competition where a contract has recently been let for an identical or similar requirement and where the value of the goods or services are small or the timescales urgent. In instances where we have an urgent need to meet additional capability requirements for specific operations, we will utilise our UOR procedure. This is used for the rapid purchase of new or additional equipment, or for an enhancement or essential modification to existing equipment. This may require existing DA leadership in order to support a current or imminent military operation or operational emergency.

Supplier selection

A7.39   In circumstances where suppliers have been selected against recently established value for money benchmarks, we will normally place additional, related requirements with that supplier.

A7.40   For UORs, rapid response is of the essence and the assumption is that there is a supplier of an available equipment that meets the need. Although not exempt from the principles of competition the rules governing advertising and bidding may be reduced or waived.

Assessing value for money

A7.41   We may negotiate with the successful tenderer, by benchmarking against the contractual terms obtained following competition, to procure the emergent requirement. In so doing we must be assured that the terms will be at least as favourable as if a separate competition had been exercised. Such practice occurs regularly for the provision of spares, consumables and services.

Greater openness: increasing the transparency of our future plans

Making a difference through greater openness

A8.1 Central to the DIS is a recognition of the need to develop much closer relationships with our industrial suppliers, with a view to promoting closer working, greater trust, increased partnership and a sense of mutual endeavour. Critical to this – both as an outcome and as an enabler to change – is the need for both ourselves and our suppliers to increase the transparency of our future plans.

A8.2 The need to respond to shifts of emphasis and priority, in accordance with changes in the global security environment, evolving costings within the programme, and to any overall changes driven by the spending review process and other planning processes, will place limits on the certainty of these plans. However we recognise that greater transparency - across the lifecycle of a capability - has the potential to offer significant benefits for both industry and Defence, providing it is appropriately managed. Being more open about our future acquisition plans will help industry to plan for the longer term and to make better informed investment decisions, with corresponding mutual benefits through-life. It will provide industry with an authoritative source of information; and it should also increase investor confidence in defence projects. We are committed, therefore, to moving substantially in the direction of greater openness with our suppliers.

What can industry expect?

A8.3 In this publication, we are already bringing together more information about our plans in a single authoritative document than ever before. We will in future continue to publish information routinely on the content of the Equipment Plan, including a catalogue listing for post-Main Gate projects of current forecast cost and ISD data. This will be regularly updated and published in the Departmental Investment Strategy, which will also include details of the MOD's non-equipment investment.

A8.4 We recognise the significant disadvantages that lack of information for industry can imply, and the advantages in terms, for example, of cost-base/overhead reduction, increased readiness and reduced cost of capital through better investment that increased openness would facilitate. If an industry decides to exit a business when we had undisclosed plans for future projects there, or alternatively invests in the expectation of a likely requirement when the most we might have is a capability aspiration, then neither industry nor MOD benefits and we may lose important sovereign capabilities. To square this circle, we are now ready to share more information than hitherto, recognising that for certain data we will only do this in a controlled environment, through nominated intermediaries empowered to speak authoritatively to industry, and with industry representatives who possess appropriate security clearances and who can protect the information appropriately. Our approach will, of course, be driven by the imperative of securing an improvement in overall value for money across the lifecycle of our capability requirements. This will necessarily condition what, when and how we communicate to industry. And we shall still need to protect information whose disclosure could harm national security or international relations, undermine internal policy formulation or prejudice our position in negotiations with suppliers.

Staff at work in MOD Main Building, Whitehall.

Sector-level information

A8.5 At sector-level, we will share information on the overall indicative planning assumptions for each sector, including illustrative Equipment Plan expenditure, as well as that on research and through-life logistics support, out to ten years. This will be at a level of detail that will enable our suppliers to make informed decisions. We will also indicate the types of technologies that we anticipate we will need to support new capabilities or the incremental growth of equipment in-service. We will also engage in an ongoing dialogue about the extent to which we believe a particular sector or business is dependent on our future orders to retain the capability and capacity needed to meet our sovereign industrial capability requirements. We will also, as far as security restrictions allow, be prepared to discuss our priorities for improved military capabilities over the next twenty years. All of this information will be shared on the basis of a clear understanding that not all of these areas will be, or may ever be, funded within the programme.

A8.6 As set out in chapter A2, our indicative planning assumptions are subjects to change at regular intervals due to government planning processes. Departmental budgets are usually set for the spending review period, and beyond that, budgets can go down as well as up. This, together with costing or strategic changes, may alter the information in the DIS.

Programme and project-specific information

A8.7 The data shared with suppliers on particular programmes and projects will vary depending on the procurement strategy we plan to follow. Where we are satisfied that there is a competitive market place and we conclude that our indicative funding allocation is adequate we will not normally release any additional financial information. If we have doubts about the affordability of our requirements we are likely to give an indication of the indicative funding allocation we have made and, where relevant, the profile of that provision. In a non competitive environment the data we release will depend on our knowledge of what the supplier can offer and the readiness of both parties to share information for mutual benefit. Other considerations will include the position of a project in its lifecycle; potential sensitivities of international partners when the project is a collaborative one; and the nature of the relationship already established with specific companies.

A8.8   Depending on how these factors apply to a particular project, the types of information that we will be ready to consider sharing with industry are:

- capability requirements and planning assumptions on production quantities;
- overall project timescales (including project in and out of service dates);
- specific planned dates for inviting and receiving proposals and tenders from industry;
- overall budgetary assumptions for the through-life capability requirement;
- planned expenditure profiles;
- associated procurement strategy;
- logistic data required to support the formulation of cost-effective support solutions; and
- how wider factors will be included in our assessment of value for money.

A8.9   In determining what information to share, our emphasis will be on releasing data that helps us to achieve our mutual objectives.

How do we intend to do this?

A8.10   Figure A8(i) illustrates the sector-level and project-specific data that may be shared with industry for mutual benefit. It sets out:

- what information may be communicated;
- who will be the principal focal point for such communications in the MOD; and
- who we anticipate will be the principal focal point for such communications in industry.

A8.11   It shows how a project's position in the CADMID cycle affects the type of information that we may expose on a particular project. Other factors - as identified above - may, on a case-by-case basis, increase or reduce the amount of information we share for specific projects depending on our judgement as to how this helps us to deliver long-term value for defence.

A8.12   In the case of sector-level information, given that our budget tends only to experience major adjustments as part of the regular planning round, it makes sense to convene a suite of 'sector days' coincident with the conclusion of our planning process. These events will provide the mechanism for presenting our indicative planning assumptions and capability priorities in each capability area (including relevant DLO spend) to our suppliers. This process will be overseen by the Commercial Director of the MOD, whom we are currently recruiting, and will bring together the relevant Key Supplier Representatives, Director of Equipment Capability, the MOD's Supplier Relations Group, relevant IPTs (wherever there is a particular interest) and Science, Innovation and Technology staff. The National Defence Industries Council Research and the Technology Committee will continue to provide a forum for MOD, DTI and industry discussions on technology issues.

Figure A8(i).

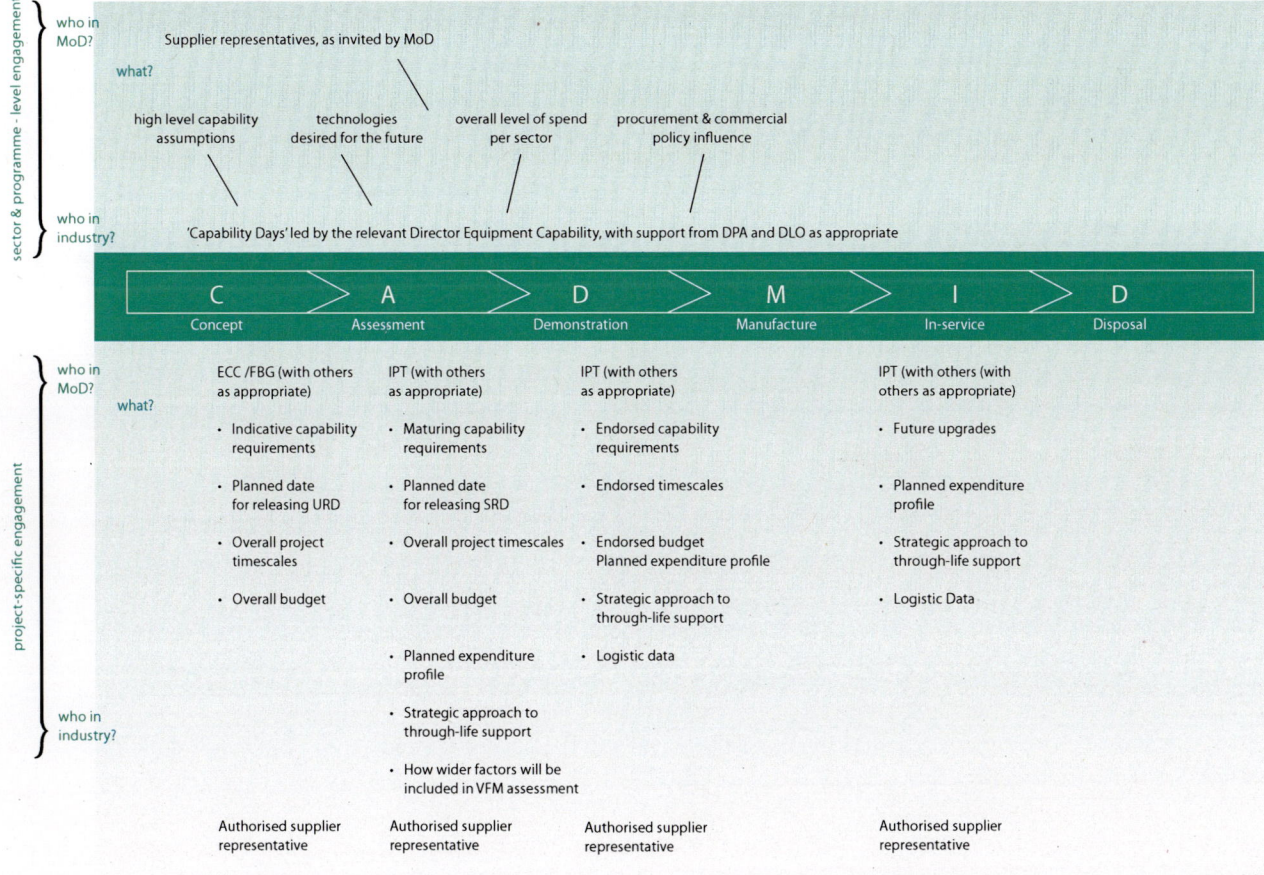

A8.13   In areas where we recognise there are particular sustainment or efficiency issues to be tackled, we will continue to consider establishing dedicated joint teams with specific companies, either on an individual or collective basis, to analyse the issues and chart a way ahead. Within such teams, with non-disclosure arrangements agreed, both sides will in general share significantly more information about MOD's plans, industry's cost base, and business models and agree assumptions for assessing potential implications. We will however only set up such arrangements where the need is clear and we see a mutual advantage; there will not be standing teams of this nature for all sectors or businesses.

What we expect from industry in return

A8.14   In return for greater visibility of our future plans, we will expect our suppliers to increase the transparency of their future plans and business information so that we can help confirm the overall coherence of what together we are aiming to achieve. In doing so, we acknowledge the need to protect the commercial confidentiality of company data. We also expect industry to acknowledge that the planning data we provide is just that. It will be offered on a 'without commitment' basis and comes with a 'health warning': to reflect changes to the environment within which our Armed Forces operate, our plans (like those in the private sector) will vary from year to year, and firm budgets cannot be set beyond the spending review period - MOD's plans are subject to change at regular intervals. Industry will also need to respect the confidence in which planning data is provided to them, in line with the obligations and expectations that are already well-established between us and our suppliers for handling sensitive data, and to be much sharper in focusing its dealings with us via authorised communication channels, rather than the much broader approach that is often adopted. In all of this, there is an explicit recognition that the greater trust that we will be investing in industry must be reciprocated: how industry responds will condition our approach in the future.

A9.1   The Defence Industrial Policy (DIP) was founded on the importance of equipping our Armed Forces efficiently with the tools they require to meet the challenges they face. This remains the fundamental starting point for the DIS. Each year a large proportion of the defence budget is spent on procuring and supporting equipment. It is these activities that in effect define and delimit our industrial policy. Above all, our relationship with the defence industry must be rooted in project performance – ensuring that reliable and supportable equipment is developed and delivered within time and price constraints. It is not just the magnitude of Government defence spending but the efficient use of those defence resources that enables the UK to have the most effective armed forces in Europe. It is not in the interests of the taxpayer or our Armed Forces for an industrial policy to dilute this fundamental principle. We also need to recognise, however, not only individual project priorities, but the complete interaction between Government and industry. And we need to be aware of the cumulative impact of potential decisions about individual projects on industry.

A9.2   It is firmly within this context that the Government seeks to maximise economic benefit to the UK from defence expenditure, by the development of technological skills, the creation of intellectual property, and an increase in the investment in the UK industry derived from exports. The Government recognises the contribution that a vibrant and innovative defence industry makes to employment, the economy and the national science and technology base. In particular, we aim to sustain an environment which will enhance the competitiveness of industry.

A9.3   This DIS covers three areas in which we can promote these goals:

●   **making the UK business environment attractive for investors to invest in our defence industry** (which we continue to define as in the DIP – i.e. embracing all defence suppliers that create value, employment, technology or intellectual assets in the UK, including both UK and foreign-owned companies). Chapter A4 explains how the Government, as a supporter, regulator, planner, customer and investor, is pursuing this. Some of the measures include, but are not limited to, the defence industry; for instance, we have specific strategies for manufacturing and technology, which include the defence industry, and the UK's stable macroeconomic environment is attractive to investors in all sectors;

●   **clearly identifying those industrial capabilities which we need, for national security reasons, to retain in the UK, and, where these are threatened, developing sustainment strategies to foster and maintain them**. Part B identifies these by sector, in the context of the future Defence requirements we seek to fill, as covered in broad terms in Chapter A2, and the changing industrial landscape, as described in Chapter A3 (both of which are covered by sector in more detail in each sectoral chapter);

●   **for specific procurement decisions, including designing procurement strategies and setting assessment criteria for competitions, ensuring that our consideration is based on long-term value for money, operational effectiveness and affordability, and also takes into account wider factors where these are relevant**. Chapter A7 gives a broad overview of how procurement strategies are designed; this chapter considers in more detail how those wider factors are identified and taken into account.

A9.4   The DIP laid out the **key factors** to be taken into account in acquisition decisions:

●   **operational effectiveness** and **cost**, whole-life not just initial acquisition cost, to assess **value for money**. The tool for assessing this is usually a Combined Operational Effectiveness and Investment Appraisal (COEIA), combining operational analysis and standard investment appraisal techniques;

●   our policy is clear that **value for money** is considered over the **long-term** and wider than that for individual projects, and in particular that any decision that would impact on our ability to compete future requirements needs to be considered very carefully. It also emphasises that there are a number of capabilities which for national security reasons we would place a high priority on retaining within the UK industrial base. The DIS builds on that policy by setting out in detail the industrial capabilities we wish to maintain, and clarifies that these two issues are often linked; any decision which fails to sustain a desired capability in the UK may affect both long-term value for money and national security.

A9.5   It also specified the **wider factors** be taken into account, which are declared and explained to potential bidders at the earliest opportunity. The critical issue is that we carefully consider the potential impact on wider national objectives before inviting industry to tender or invest significantly in a project (otherwise over time, industry's confidence that our decisions will be fair and transparent will decline, with knock-on consequences for value for money and military capability). Consideration of wider factors applies equally in cases where prime contractors run competitions on behalf of the MOD.

Security of supply

A9.6   We need to ensure that we can support equipment, or produce expendables (e.g. munitions), in times of conflict (predicated on an assumption that we understand the dependencies within the supply chain, where in some cases we need to do further work with industry). High levels of onshore technology and capacity may often offer greater comfort in security of supply and the ability to undertake modifications in response to short-term operational demand. This applies as much to support as to initial procurement.

A9.7   The DIP emphasises that we need to be realistic about security of supply advantages, recognising that increasing mutual reliance on security of supply is inevitable for all nations. The weight to be put on security of supply is a question of judgement case by case, taking into account the risks involved (including any mitigation provided through collaborative agreements such as the six nation European Letter of Intent Framework Agreement and the US/UK Declaration of Principles Security of Supply arrangement) and the cost implications. In some cases, the capability is so significant to our overall military effectiveness that we wish to retain it onshore at least partly for that reason; security of supply, in terms of both physical and intellectual resources, is often a critical factor in deciding which capabilities must be sovereign. In most other cases, however, we will need to balance the risk against any additional cost for onshore supply on a case-by-case basis, taking into account value for money and affordability. In inviting industry to bid against such requirements, we need to avoid a 'UK premium' being priced into domestic bids.

Industrial participation

A9.8   The MOD's Industrial Participation (IP) policy can encourage technology transfer and ensure investment in particular industrial capabilities within the UK. Where this contributes to developing or maintaining sovereign capabilities, IP may be a key factor in its own right; where the benefit is of broader industrial benefit, it may help discriminate between two otherwise similar propositions. If however the benefit is a broader industrial one, but securing it would involve selecting an option which was worse for Defence, then it would need to be considered in the same way as other wider technology and industrial capability factors. It is important that MOD has confidence in a foreign supplier's ability to deliver the IP offered, so factors including the company's past record will be evaluated.

Industrial capabilities

A9.9   We continue to recognise that there are industrial capabilities which do not meet the strict defence criteria for sustainment but may be desirable to retain in the UK due to the high value they bring to the industrial economy, wherever they sit in the supply chain. Their significance can be evaluated according to:

- their potential in world markets where the UK may gain greater market share;

- the extent to which they will generate high-value added economic activity (including its potential for attracting inward investment and incorporation into collaborative programmes);

- transferability into wider commercial applications outside defence sectors;

- impact on industrial activity regionally (including the number and quality of UK-based jobs that are created or sustained).

A9.10   All business cases on equipment (including support business cases) are required to have supporting sections considering industrial and regional implications. Assessment of such factors must however recognise that funding for such capabilities would be drawn from the defence budget.

Key Technologies

A9.11   In some cases, key underpinning technologies are important to maintain for national security reasons, and a number are identified in Part B. MOD's science and technology strategy necessarily focuses the limited research funds available on investment in technologies judged to be of most importance for defence. These technologies will often have potential for the wider UK science base, a base also supported through DTI science, innovation and technology programmes, and will change over time, sometimes rapidly. Acknowledging the relationship between military and civilian priorities and investment strategies, in part through an ongoing engagement between MOD and DTI, is important both for MOD's own Technology Strategy and for formulation of the MOD research programme and may also help identify areas for coordination and collaboration. The interaction between military and civilian priorities may also in some cases be relevant in other acquisition decisions.

Export potential

A9.12   Defence exports help to maintain key industrial capabilities during lulls in domestic demand allowing longer production runs and reduce the share of fixed overheads on MOD programmes. They also contribute to foreign and security policy interests. These benefits are discussed further in chapter A6. Exports can also improve the economic strength of the defence industry. For theses reasons, a realistic assessment of export potential, and the benefits that might accrue (to Defence or more widely) is made at the key decision points in MOD programmes.

Foreign and security policy interests

A9.13   Many procurement projects and collaborative ventures can be so large in scale and political importance that they have significant implications for these interests. Choosing whether to cooperate with a particular country in a joint programme could have consequences for the overall bilateral relationship and an alliance's military capability. For example, agreeing to cooperate in a joint programme could strengthen another country's political commitment to their own military's modernisation, which we would value as part of our broader security policy because of the greater burden-sharing that might then be possible in future combined operations. It is usually very difficult to quantify foreign and security policy interests objectively and the potential application of this heading is very wide, but they need to be assessed and considered carefully. Given the relationship between defence and broader security and foreign policy, a decision which is suboptimal in a narrow consideration of the key factors may nevertheless be the best for Defence in the round. However, the interests concerned must not be transient or trivial.

Wider MOD policy

A9.14   This covers a broad range of policy areas in which MOD Ministers, or for some matters relating to internal and Armed Forces administration the Defence Management Board, have chosen to constrain the Department's actions (legal obligations are of course absolute, and all business cases must be considered for legality of the equipment including in the way we intend that equipment to be used, but policy choices are more discretionary). For example, to comply with MOD environmental policy, we have specified that future tanker shipping should be double-hulled, and for safety reasons, that future munitions should be insensitive. Given knowledge of existing policy, options will rarely be offered which would breach Departmental policy, except where compliance has a significant impact on the key factors. Options that required exclusions from existing policy would need discussion with the policy lead for that area, and reference to Ministers and senior officials as necessary.

section B

Review by Industrial Sector and Cross-cutting Capabilities

Systems Engineering
Maritime
Armoured Fighting Vehicles
Fixed wing including UAVs
Helicopters
General Munitions
Complex Weapons
C4ISTAR
CBRN Force Protection
Counter Terrorism
Technology priorities to enable defence capability
Test and Evaluation

B

Review by Industrial Sector and Cross-cutting Capabilities

The charts in this section are a snapshot of MOD's current internal planning assumptions for the indicative spend in each sector. As noted in section A8 they are subject to periodic review and over time may go down as well as up. The resources available for individual sectors will be influenced over time by the overall size of the defence budget, the relative priority given in the MOD planning process to different areas of defence capability, and the need to flex specific plans if necessary to address cost pressures in other parts of the defence budget.

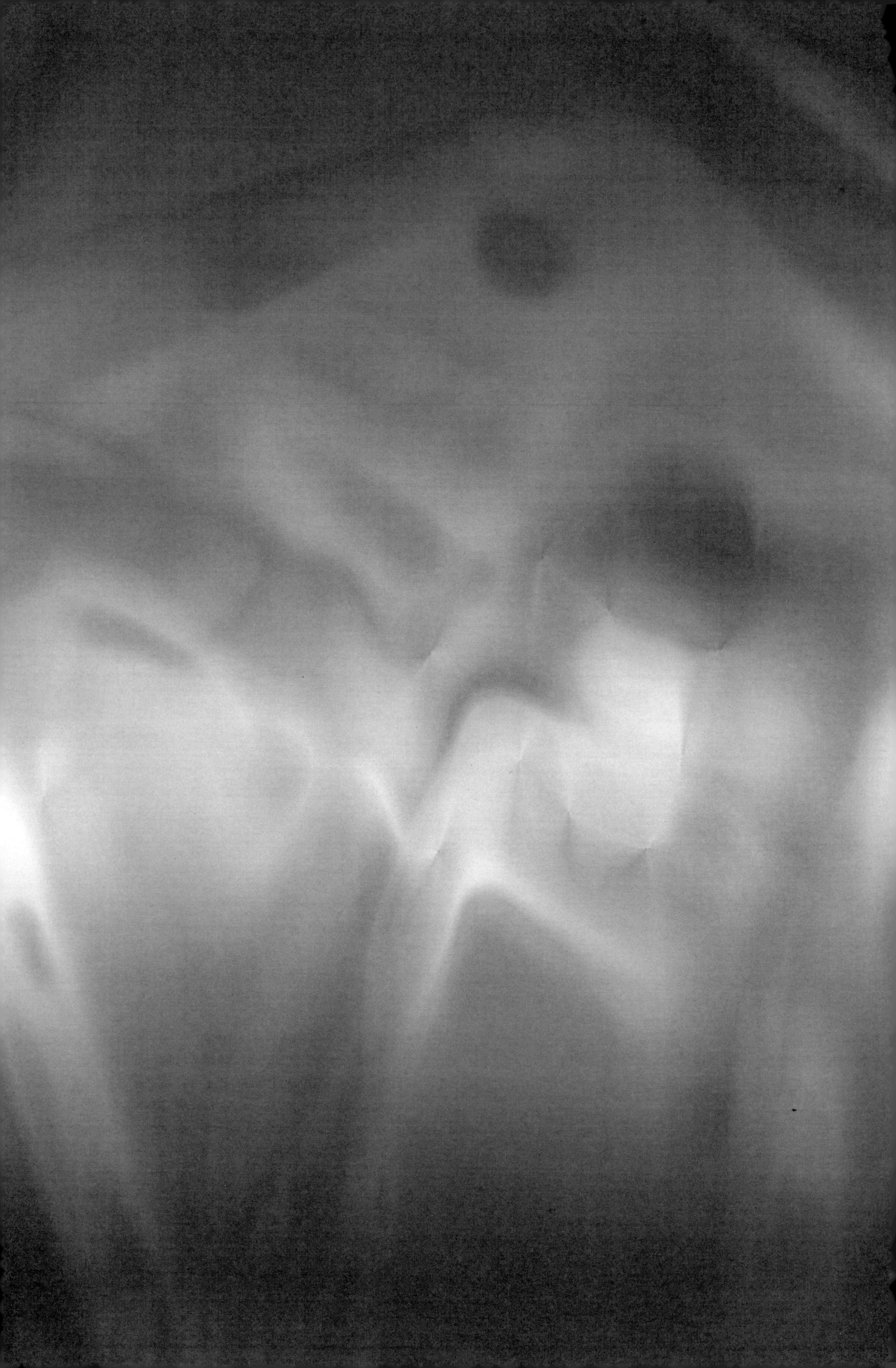

Systems Engineering

B1.1   The DIS has been developed using Key Principles, explained in detail in Chapter A1. One of these was recognising the importance of systems engineering, to ensure that our equipment is procured and continues to develop through-life as a result of mature discussions between intelligent customers and intelligent suppliers.

B1.2   In general, we need industry to have systems engineering capability so that the UK can:

- integrate systems of systems;

- design systems and upgrade them efficiently through-life;

- integrate sub-systems or components, in some cases sourced from around the world, into systems;

- provide a core capability from which to surge when demand requires, e.g. to address specific requirements for a new operation or to provide a planned upgrade, and as new technologies emerge;

- adapt systems to take advantage of new technology and to respond to changes in the threats.

B1.3   Along with industry, we need to invest in maintaining and growing high quality systems engineering capability, at all levels in the supply chain where we need systems or key sub-systems to be designed and engineered. But the level of capability that we require onshore varies by sector.

B1.4   **In a period when platforms are likely to remain in-service for many years, unless systems engineering capability and vital long-term knowledge is maintained, it is little use investing in cutting-edge science. New technologies will have less benefit without knowledge of how they might be exploited and inserted into existing equipment.**

The late US Admiral Grace Hopper, the eminent computer scientist, educator and thinker who first coined the term 'bug' for a programming error, once said, 'Life was simple before World War II. After that, we had systems'. Despite systems engineering as a discipline having begun its development around sixty years ago, its scope and definition remain subject to debate even amongst qualified practitioners, despite attempts by the International Council On Systems Engineering (INCOSE) to establish a 'Consensus' (http://www.incose.org/practice/fellowsconsensus.aspx). Within industry, different companies (and different systems engineers within the same company!) have different views on the scope of what is meant by terms such as Systems Engineering and Systems Integration, and the relationship between these and e.g. project management. This chapter does not aim to create a new set of definitions, but to explain what we mean by systems engineering in the Defence Industrial context, and why it is important.

B1.5   As military effect is increasingly delivered through the interaction of many different people, equipments and information systems, all of which will develop over time, a clear understanding of these evolving relationships from the point of a potential need being recognised, through development and manufacture, to the end of the solution's service life is essential if its design is to be balanced and high performance.

What is systems engineering?

B1.6   Modern defence equipment and services delivered using equipment generally rely on discrete elements, often complex in their own right, relating to each other in a planned and well understood way. This combination of different elements, delivering an overall result which is greater than the sum of its parts, is a system, and military capability is delivered through systems.

B1.7   Systems engineering is the general term for the methods used to provide optimally engineered, operationally effective, complex systems. Systems engineering balances capability, risk, complexity, cost and technological choices to provide a solution which best meets the customer's needs.

B1.8   As defined by the International Council On Systems Engineering (INCOSE): 'Systems Engineering is an engineering discipline whose responsibility is creating and executing an interdisciplinary process to ensure that the customer and stakeholder's needs are satisfied in a high quality, trustworthy, cost efficient and schedule compliant manner throughout a system's entire lifecycle.'[1]

B1.9   The through-life relevance is important; even at disposal, there may be system safety or security issues to be considered, or we may have to modify equipment to meet the needs of an overseas buyer.

Why does system engineering matter? Initial considerations.

B1.10   Together with the defence industry we are in the business of delivering large, complex projects, often at the forefront of technology. The National Audit Office has documented the major sources of difficulty associated with the delivery of our major projects, the majority of which are associated with technical issues. Many of these technical issues relate to the integration of the systems involved. Improving systems engineering is, across the industry and MOD, a high priority if the Armed Forces are to get the equipment they need.

---

[1] *Systems Integration is another term sometimes used synonymously with Systems Engineering; as the ICOSE consensus (see box) describes integration as a discrete activity within a Systems Engineering process, we will generally avoid that term to avoid confusion.*

B1.11   Individual acquisitions are increasing in their complexity, as technology develops and as military effects increasingly are delivered through a combination of different platforms, forces and information systems.   Realising Network Enabled Capability (NEC) is all about making different capabilities work together in a coherent system, to deliver a step change in capability. As our own systems become more complex, achieving interoperability with our allies becomes an ever greater challenge.   Open architectures can help manage this trend, but the underlying drivers remain. Our priorities for our forces include flexibility, precision, agility, and reach. Good systems engineering (particularly in pursuit of sustainability) can help design out unnecessary complexity, but in general, a flexible, precise, agile, and long-range capability, will be a complex one.

B1.12   Maintaining capability through a system's life, often longer than any individual's career, requires the original understanding of the system to be retained, with the basic rationale for previous trade-offs, and the dynamics of the relationship of the system's parts, captured and understood. Only by doing this can the implications of integrating new equipment be understood, and opportunities seen for inserting previously unavailable technology to improve the system's safety and performance or drive out cost.

Systems engineering and prime contractors

It is very important not to confuse the concept of a systems engineering entity with that of **prime contractor.**

A prime contractor is the legal entity whom we require to deliver a product, e.g. a fully functioning ship, or service. A systems engineering capability requires high calibre engineering skills, a suite of modern organisational, modelling and simulation processes and tools, and access to facilities to test and prove the system in all aspects of its operation. Firms acting as prime contractors will often possess much of this capability, but there is no necessary reason why the prime contractor for a specific equipment or service should not subcontract some or all systems engineering tasks to another firm or group of firms, or why they should have to own all the relevant facilities, provided they have assured access to them. The optimal relationship between prime contractors and systems engineers is considered further below.

B1.13   The complexity of the systems engineering tasks necessary to deliver military capability differs by sector, with a nuclear-powered submarine perhaps representing the largest, most complex and highly integrated platform system, compared to a relatively simple Armoured Fighting Vehicle (AFV), with more limited systems complexity. **The trend is nevertheless clearly towards greater systems complexity across all sectors**, and the latest AFVs may now include a range of interdependent sensors and electronic sub-systems, and a huge number of lines of software code.

Different levels of systems engineering

B1.14   Ultimately, we have to be able to bring together our resources, information and the operation of different force elements to deliver military effects. To ensure the desired effects are available, planners have to take into account the Defence Lines of Development:

- Training;
- Equipment;
- Personnel;
- Information;

- Concepts and Doctrine;
- Organisation;
- Infrastructure;
- Logistics.

B1.15   The provision of the equipment capability is only one element therefore of military capability, and the development and operation of military capability is a system in its own right. Those charged with delivering the equipment element have to ensure that it is consistent with what the other Lines of Development can deliver, in a dynamic relationship. This coordination is part of our core business, but those designing and developing equipment capability have to take them into account, and understand the customer's needs, constraints, and intentions.

B1.16   Systems engineering typologies can be drawn up concentrating on, for instance, operational, industrial, business, project, or technical aspects. Given the different perspectives possible, systems engineering is as relevant in designing a software chip as it is for considering development of the military strategy for a particular conflict or the future force structure.

B1.17   Equipment therefore needs to fit into a broader system with the other Lines of Development, and different perspectives are possible (e.g. an aircraft can be considered a system in its own right, or part of the system-of-systems that is its squadron, including ground crew, supply chain, information network etc).

B1.18   Nevertheless, significant equipment programmes generally represent complex systems in their own right. A number of different perspectives can be taken, and it is traditional to talk about 'Tier 1' or 'prime systems integration' systems – which tend to be platforms – and 'Tier 2', which are (often still complex) systems incorporated into platforms. This is generally illustrated by a 'V-diagram' like the one opposite:

System Engineering "V" Diagram

B1.19   One definition of Tier 1 is as '**the level of major systems that support key defence capabilities, and which are supplied directly to the military user (e.g. a military aircraft, helicopter, warship, vehicle, guided weapon, satellite, standalone sensor or C3I system)**'. However, this makes clear that this typology is descriptive rather than definitive; systems are put into tiers based on how we have traditionally contracted for a system.

B1.20   This risks confusion; e.g. guided weapons are clearly complex, and are generally contracted for separately from the platform(s) they reside in, but they do not represent military capability until incorporated onto those platforms - and usually, into an information network, either via the platform or directly. Nevertheless, the defence industry still tends to be structured around 'Tier 1' and 'Tier 2' products. Below Tier 2, there may be decreasing levels of complexity or less specific military features, until the item being used is effectively a commodity raw material.

B1.21   In the remainder of this chapter, to establish a consistent typology and to avoid potential confusion between what is a Tier 1 and Tier 2 system, we distinguish between:

- **partial systems**: these have independent purpose, but are only viable in the context of a containing system. For instance, an air-launched missile has an independent purpose, but is only viable if integrated with an aircraft platform;

- **sub-systems**: systems which only have a purpose as part of a containing system. For example, an engine only has a purpose – providing the power to move a platform – if incorporated into that platform;

- a **system**: one which has a purpose and is viable in its own right. An example, from an equipment perspective, is a fast-jet combat aircraft. We will generally talk about **platform systems** – in the military sense, as the single viable equipment units, usually capable of independent movement – though examples include satellites, as well as vehicles, aircraft, ships and submarines. **Network systems** also exist, however; a Wide Area Network fits this definition of a system;

- a **system of systems (SOS)**: these contain systems which have purpose and are viable independent of the SOS, but which can when acting together perform functions unachievable by the individual systems acting alone. For instance, the future aircraft carrier, combining its aircraft carrier group with its own sensors, communications and command systems and weaponry and interacting with wider networks, represents a SOS[2].

B1.22   The growing importance of networks and their interaction with partial systems in particular may make a platform-centric perspective less useful in future when considering how to meet operational capability requirements, and where the critical interfaces between systems may reside. Some systems, particularly those which seek to integrate a number of different sensors and weapons systems across platforms, are likely to require deeper and more complex integration into their platforms <u>and</u> into networks. This may sometimes require deep knowledge of the sub-systems involved and their potential contribution to military capability, separate from their physical integration into a platform system.

One rule of thumb used in industry is that 'a good systems engineer understands the details that are being addressed at least two levels below him and one level above him'

B1.23   Engineers of platform systems do not always need a deep level of understanding of the partial systems or sub-systems; it depends on the criticality of the partial or sub-system concerned and the system architecture being used. If the architecture has clearly defined interfaces and allows a 'plug and play' approach, and understanding the underlying technology is not critical to confidence in overall

performance, then this may be less important. Similarly, engineers of partial systems and sub-systems may not always require deep understanding of the overall system, though if they have this they can contribute to the overall challenge of optimising the system.

Partial or sub-systems can be highly complex in themselves. For instance, the Rolls-Royce EJ200 turbofan engine pictured below was designed by selecting European centres of excellence in gas turbine subsystem design and whole engine integration. Rolls-Royce integrates whole engine performance, utilising the expertise of four leading engine manufacturers sharing their technical responsibilities and know how to ensure the EJ200's competitiveness. It includes advanced features such as:

- a fan design that gives high stability without the need for Inlet Guide Vanes;
- blisks (a one-piece disk-and-blades engineered from a single piece of material);
- wide-chord aerofoils;
- single crystal blades;
- an airspray combustion system;
- an integral Full Authority Digital Engine Control.

This photograph is reproduced with the permission of Rolls-Royce plc, copyright © Rolls-Royce plc 2005.

Partial and sub-systems can also be major contributors to the military capability sought: a guided weapon, for instance, fails or succeeds depending on the sophistication of its sensors.

B1.24   However, at each level of the process, <u>some</u> degree of understanding of the systems engineering task at the next tier above and below is generally implied, even if it is only to act as an intelligent customer and be fully aware of the problem to be solved and all the potential solutions. Although many of the tools and techniques of systems engineering are generic, their application will often need specific expertise in the relevant sector.

---

[2] It is of course possible to see the carrier, from an operational perspective, as itself part of a wider system of systems, e.g. the carrier battlegroup, or indeed the complete set of defence resources that can be configured to a greater or lesser extent on demand to meet the changing needs of Government policy.

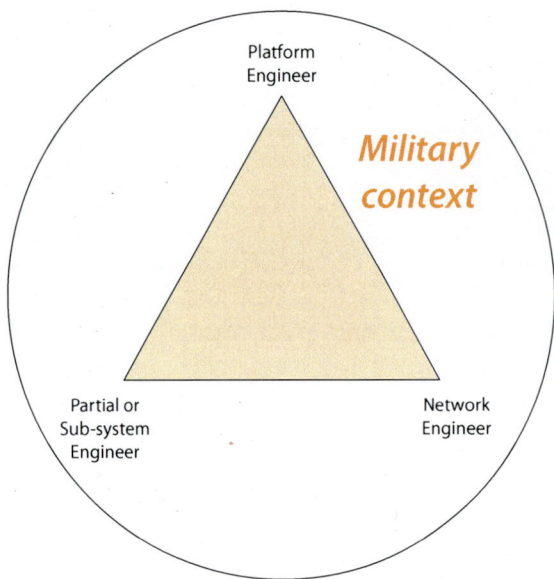

Military context

- Platform Engineer
- Partial or Sub-system Engineer
- Network Engineer

B1.25 We now expect our platforms to endure in-service for long periods, supported by both regular and operation-specific upgrades. At the same time, technology, particularly in electronics, continues to evolve rapidly. This may imply that the most significant opportunities for upgrading capability will be identified and resolved at a) the systems of systems, or b) the partial systems and sub-systems level, rather than being driven by fundamental modifications to a platform architecture. Maintaining knowledge of the system as a whole, and a Design Authority, is a crucial enabler to this – if the implications of modifying a sub-system for the system as a whole are unknown, it will not be safe to incorporate the change, and the platform systems engineer may remain best placed to identify areas for investigation to drive out cost and improve availability. But the innovation and ability to exploit technology development and improve other aspects of military capability may increasingly reside at these lower, partial system and sub-system, levels. This is why the model places all the different dimensions within the context of military capability; those with the relevant engineering knowledge in all domains also have to understand the operational challenges, if they are to be able to identify opportunities and the highest priorities for investment.

B1.26 The recognition of the importance of some sub-systems and partial systems also means that once equipment is in service, notwithstanding the potentially closer relationship between the engineer of the platform system (who has traditionally been the prime contractor for the capability) and the MOD, both have a clear interest in managing the supply chain very closely. This is not only to drive out unnecessary cost, but also to ensure key underlying industrial capabilities are maintained, and sub-system/ partial systems engineers share the motivation to improve equipment through-life. This is not achieved if the engineer of the platform system or the prime contractor can exercise inappropriate dominant power. Rather, it implies that, where contracting arrangements can be suitably designed, the platform engineer should only be rewarded for the value that it adds (including, if the prime contractor, in managing the supply chain), with innovation rewarded where it arises, so partial systems and sub-systems engineers share the incentives for continuous improvement.

B1.27 The capability to engineer at the platform systems level does not, therefore, of itself imply a vertically integrated supply chain. To ensure all avenues of innovation are available, vertical integration could sometimes be counter-productive.

- The systems engineering challenge in Defence equipment is increasing, as platforms have longer planned in-service lives, but technology, especially electronics, continues to evolve rapidly.

- The complexity varies by sector, from those where the overall performance of a single platform in a demanding environment (e.g. underwater, air) is the critical function, to those where the platform is a relatively simple host for a range of partial systems and sub-systems which deliver a variety of military effects.

- In most cases, the key opportunities for technology insertion are likely to come at the partial systems/sub-systems layer, but these can only be investigated if the systems engineering knowledge of the overall system architecture is maintained.

- Depending on the likely sources of capability-advancing innovation, partial systems and sub-systems engineers need a good understanding of the overall military context and the problems the military customer is trying to address. It is in MOD's interest to promote this. They also need to share the incentives for continuous improvement.

- Platforms and their partial systems and sub-systems increasingly have to operate in a networked environment, where systems of systems are reconfigured to deal with new opportunities, often at short notice. This creates challenges by potentially changing the military context for a particular system, but the added complexity at the technical level can be mitigated by the use of open architectures and common standards.

The strategic importance of platform systems engineering

B1.28 In defence, even 'commodity'-like items can require deep systems engineering capability to deliver; for instance, the design and production of current and new generation microchips is a highly complex task. A deep understanding of commodity engineering is not, however, a general requirement. The UK also has a number of partial systems and sub-systems suppliers with successful relationships with international platform systems suppliers and who, on the strength of their technology and engineering capability, compete effectively in the global market.

B1.29 There are in contrast relatively few defence companies with significant proven platform systems level capability. The importance of systems engineering at this level is not just as an effective and necessary discipline to ensure programmes are delivered to time, cost and quality. In many areas, it is fundamental to delivering the other Key Principles of the DIS. The platform, after all, represents the physical, viable unit, within which all the partial systems and sub-systems come together.

Appropriate sovereignty.

- **Operational independence and being an intelligent customer.** Systems engineering capability is central to understanding whether the system will operate as you want it to, when delivered, and as it evolves through life; it may not always be possible to tell this simply by independent testing. This applies both for initial purchase and for support and upgrades. Having reliable access to this capability within the UK, particularly for Urgent Operational Requirements, is generally a high priority.

- **Avoiding the 'captive customer' risk.** Relying on an overseas platform systems engineer could limit the ability to develop and upgrade equipment to meet unique UK requirements, unless there are credible and clear contractual and political guarantees. In some areas, we may be prepared to be share sovereignty here[3]. But in the worst case, reliance on an overseas systems engineer could lock us into having to agree to inappropriate, unwanted and expensive changes in configuration, or risk the systems engineer withdrawing support from older variants.

- **Strategic industrial influence.** Without an onshore candidate platform systems engineer, our negotiating leverage in procuring equipment competitively in the global market would be markedly reduced and we could be exposed to overseas monopolies. And in cooperative programmes, it is important to be able to participate meaningfully on an equal or near-equal footing with international partners.

- **National provision.** In some areas, overseas sourcing is impossible, for legal or security reasons. The ability to develop such systems <u>has</u> to be maintained on shore.

B1.30 Given all these considerations, maintaining a UK systems engineering capability in defence sectors has a broader political and strategic impact: it signifies the UK's status as a major defence nation; it allows the UK to bring to coalition operations unique or distinctive capabilities; and in some areas like nuclear submarines, it allows the UK to produce strategically significant, complex systems which are not available, or which we would not wish to source, from the international market.

B1.31 **Through life capability management** cannot be effectively delivered without a platform systems engineer that understands not only the systems architecture, but the reasons it developed as it did. The necessary knowledge may be lost at:

- individual level (retirements, resignations, deaths and the normal process of forgetting, or not realising what is important),
- collective level (through the break-up of established teams, the outsourcing of functional areas, or a change in ownership of part or all of a business),
- or, even when the knowledge has been recorded in a suitable form, through the loss of archived knowledge (actual loss, accidental destruction, physical deterioration, or being captured on obsolete media).

B1.32 Addressing these risks requires the importance of the knowledge to be recognised, and MOD and industry to set the right incentives for its documentation, and transfer. Without this, equipment may become obsolete unnecessarily early, because no-one understands well enough how it might be modified. Designing equipment with through-life capability management in mind may even change the design adopted, as it can be designed to make future modifications easier, for instance through using open architectures and international standards.

The why-knowledge and why-not-knowledge

According to the Defence Scientific Advisory Council: 'Without strong leadership and firm goals, efforts to manage knowledge spiral down because they rarely provide enough short-term benefit. A classic example is the way engineers skimp on capturing the rationale for decisions, the "why-knowledge" and "why-not-knowledge". Without such knowledge it is often impossible to know later whether one can safely modify a system. But for the engineers at the time, recording such knowledge is a chore, and for a company the effort detracts from project delivery and profit.' Both MOD and industry need clear goals and strong leadership to maintain vital long-term knowledge.

B1.33 **Maintaining key industrial capabilities and skills** in turn depends on this knowledge being captured, refreshed, and utilised even when the next major platform of that type is not foreseen for many years.

- In the past, a steady succession of new platform systems and midlife updates maintained these; now the gaps between one platform and the next are generally much longer. MOD and industry increasingly need to work closely together to maintain this capability.

- This means aligning incentives better, and changing business models. Vital domain-specific platform systems engineering information will not be maintained if industry brings together its '**A Team**' for the initial acquisition and production but then moves this on to the next major programme; but equally, a company cannot be expected to keep its best people employed maintaining skills which we value but which do not reflect the **best short-term return available to the company**.

- However, moves in the Defence Logistic Organisation to **contract for availability** and the use of equipment rather than the production of spares, demonstrate that industry can be incentivised to maintain systems engineering knowledge, at both platform systems level (as usually embedded within the original prime contractor) and lower levels, to keep working on improving reliability to reduce cost to MOD and increase profit to their shareholders. And as industry sees that we are able to insert new capability rapidly as technology develops, it will be motivated to invest its own resources, alongside our research, to help us understand the opportunities and offer unsolicited proposals for improving our capability.

- Similarly, this proposition offers potentially interesting **career paths** for individuals; gaining deep knowledge, but investigating the application of exciting new technology into existing platform systems, as well as more traditional Post Design Services.

- This model of activity needs the **focus** an entity at platform systems level can bring, to coordinate and promote research, and to identify and plan to develop the critical technologies, teams and skills (some of which may be deep in the supply chain) to realise these commercial opportunities and maintain defence capability.

---

[3] For instance if the costs of maintaining a different UK configuration compared to others in a cooperative programme were poor value, or we were still able to purchase UK-specific change requirements from an overseas supplier.

Apache assembly.

Contracting for availability

Our operational experience highlights the importance of reliability. We are procuring generally more reliable equipment and, consequently, there is a reduced requirement for moving spare parts and replacement equipment. Less broken equipment being sent back for repair also reduces pressure on the supply chain. It also means moving towards a system whereby we pay for **availability not repairs.** In the past equipment has been procured with two separate contracts, one for delivery and one for repair. This does not clearly incentivise the delivery of reliable equipment. We are moving to a single contract system, by which suppliers are paid for **use** of equipment. A recent example of this is the £600m private finance initiative for 'C-Vehicles', under which the Amey Lex Consortium will provide heavy plant equipment, logistic support and construction machines over a 15-year period on a rapid fleet turnover basis. This will reduce maintenance and support costs.

Companies are generally keen to move towards this model. While it offers us greater equipment availability, it also provides our industrial partners greater returns over a longer period. The Chief of Defence Logistics' intention is to spread this business, where it is appropriate, across the whole support area.

B1.34   Maintaining engineering capability at the platform systems level can also provide **Value for Defence.**

- This model of an incentivised platform systems engineer working to develop a design for ease of upgrade and support should directly **improve potential equipment capability** and **reduce costs** to us.

- It may in turn also improve **export prospects**, potentially for the mutual advantage of us and industry. This issue is explored further in the supporting chapter on defence exports.

- Having a domestic capability at platform systems level also assists in securing Value for Defence in competitive global procurements. Unless the UK has this itself, it will be viewed as an 'export market'. Without the focus that a domestic platform systems-level engineering capability can provide, the **value gained from our research** is likely to decline, as might national innovation in the supply chain. This is because overseas platform

systems engineers would be concerned to use their own national resources, and overseas governments will often naturally have a policy concern to favour their own technology suppliers, not least to ensure that their industry can export without external constraint. This would leave UK SMEs in particular without an easy **route to market** for their innovation, with consequences both for Defence and the wider UK science and technology base.

Overseas systems engineers

The Defence Industrial Policy states that the UK defence industry should be defined in terms of where the technology is created, where the skills and the intellectual property reside, where jobs are created and sustained, and where the investment is made. Thus an 'overseas systems engineer' in this context is a company where the key systems engineering capabilities, including the relevant intellectual property, are outside the UK. That includes companies which might have significant presence in the UK, e.g. manufacturing facilities, and in all other senses be considered part of the UK defence industry. It would equally include primarily UK-based or co-owned companies which moved their systems engineering capability and associated IP offshore.

That does not mean that overseas companies could not offer their products in areas where we specified a requirement for a platform or other systems engineering capability in the UK, either by establishing it themselves (although the barriers to entry may be high), or by acquisition, licensing or partnership with a firm that does have such capabilities onshore. For instance, it has long been important for the MOD to be able to modify, often for Urgent Operational Requirements, its C-130K Hercules aircraft, originally bought from the United States forty years ago. This has been achieved through the Sister Design Authority[4] established by Marshall of Cambridge for the original C-130K purchase. Marshall has carried out over 1,000 upgrades, modifications and new equipment additions to the RAF's fleet and has attracted substantial further work from overseas because of its acknowledged expertise.

B1.35   **Change on both sides**. As implied in the above discussion, ensuring the retention of systems engineering capability in the UK is in general important but needs, for implementation, to be underpinned by different behaviours in both us and industry.

- In industry, amongst other things, it requires clear **leadership** and **investing in maintaining and refreshing skills and knowledge**, and a marketing mindset – helping anticipate MOD's and export customers' needs and finding suitable solutions, rather than relying on known future programmes and being told the system requirements.

---

[4]   *Sister Design Authority: A contractor appointed to operate in parallel with the original, or current Design Authority (typically an overseas organisation) for a design, but which has access to the design records and design data, and has sufficient expertise and familiarity with the design to act as the Design Authority. The Sister Design Authority is responsible for impact of the design work it carries out. E.g. for airworthiness, with its access to the original design data, a Sister Design Authority is competent to decide when the opinion of the current Design Authority should be sought on changes which might affect airworthiness. The original Design Authority remains responsible for the airworthiness of that part of the design which has not been affected by the work of the Sister Design Authority.*

- For us, it implies allowing **increased profit** for industry in conjunction with **reduced costs**, aligning with industry to support, through minor programmes and research, the maintenance and development of vital knowledge.

- It also means demonstrating that we can be **disciplined** enough not to ask for overly demanding specifications at first delivery, while being **agile and flexible** enough to be able to react when industry does produce sensible proposals for rapidly taking advantage of new technology. These challenges are considered further in Part C.

Industry overview

B1.36   We have discussed the strategic importance of systems engineering, and outlined a potential vision in which a UK systems engineering capability, particularly at platform systems level, can help sustain the Key Principles of the DIS. The vision however needs to be rooted in the real world of the industry as it currently is, and the available methods for engaging with it.

B1.37   Systems engineering is a complex task involving a mix of generalised skills and resources (including project and programme management and standardised systems engineering techniques) and domain-specific knowledge and facilities, including test and evaluation facilities which are discussed in another chapter. Very few countries outside the US are able to support more than one capability at the platform systems level in any specific sector, and even in the US the largest systems engineers still tend to concentrate on a limited number of sectors.

B1.38   The industrial overview for each sector is explained in more detail in the relevant chapter, and there are a number of companies which can contribute different capabilities: for example, BMT, QinetiQ and Three Quays have some proficiency in naval systems engineering, and naval maintenance also requires understanding the systems design. But across the piece, the domain-specific capability at platform systems level within the UK defence industry is currently primarily concentrated as follows:

Sector	Company
Fast-Jet combat aircraft and maritime patrol fixed wing	BAE Systems
Helicopter	AgustaWestland UK Eurocopter (Puma and Gazelle)
Strategic airlift (C-130)	Marshall of Cambridge
Submarines	Babcock Naval Services Ltd, BAE Systems, KBR (including DML)
Complex surface warships and Royal Fleet Auxiliary	Babcock Engineering Services Ltd, BAE Systems, KBR (including DML), Thales, VT
Armoured Fighting Vehicles	BAE Systems
Complex weapons	MBDA(UK), RSL, Thales, BAE Systems UWS
Non-embedded C4ISTAR	BAE Systems, Thales, EADS, General Dynamics, Lockheed Martin, Northrop Grumman, Raytheon, Selex Communications, VT Communications, Ultra Electronics, BT, EDS, Fujitsu, LogicaCMG, QinetiQ
CBRN	Smiths Detection, General Dynamics UK, SERCO Assurance, EDS

B1.39   It is important to emphasise that this is a view of the current state of the industry. In implementing the DIS, Government and industry will need to do more work to refine in detail the knowledge (including IP), skills, facilities and capacity required, and identify where these reside. There is no reason in principle why additional capable systems engineering capabilities at platform systems level could not emerge in the UK, for instance if an overseas supplier of a new equipment established such a capability onshore. Thus, the fact that BAE Systems has supplied 95% of the UK's current Armoured Fighting Vehicles and is the only current on shore engineer of systems in this sector does not mean that another company could not establish that capability for future projects, provided we could access the intellectual property design authority and capability to upgrade or adapt those future vehicles as required, including against demanding timescales. Equally, there is no reason in principle why the systems engineering capabilities currently embedded in these companies should have to remain where they currently are.

Sustaining the capability to engineer systems

B1.40   The proposed way forward is explained in each sector chapter. However, it is important to emphasise again that, where we say there is a national requirement to retain a capability to engineer at the systems level, whether for upgrading in-service equipment only or also for new projects, that does not of itself imply a particular model for engaging that capability. **In some sectors, it will not be necessary, possible or cost-effective to retain this capability.**

Services:

In some cases, industry may deliver some of the other Lines of Development itself, either with our assistance or without. The Strategic Sealift Service, which came into service 20 months ahead of its target date, is a good example. Strategic Sealift is about moving vehicles and equipment to a theatre of operations quickly and efficiently. Each of the six 20,000 tonne Roll on, Roll off (RoRo) ships is almost as big as an Invincible Class aircraft carrier, with three decks, each deck nearly the length of two football pitches. Designed with transporting military equipment in mind, the vessels can dock at a wide variety of ports, loading from either the stern or side. The ships also carry a 45 tonne crane, with enhanced stability and ice breaking capabilities. The ships are clearly crucial elements of the service. But industry also provides the personnel and the majority of their training itself and is responsible for generating business for the ships when not required by us, managing their activities to meet both our and commercial customers' requirements (with the former always coming first). We receive a service rather than a product. The PFI contract will run until December 2024 and the full cost of the service will be approximately £950 million, subject to usage.

The relationship between systems engineering capability and prime contractors

B1.41   We have noted throughout that being able to engineer capability at a systems level, and prime contracting, are not necessarily synonymous. Nor is there a simple model of the degree of vertical integration that is desirable. However, taking into account different levels of systems complexity and the overview above of potential methods of engaging with industry, we can identify some general principles:

● for new projects, where the platform is itself highly complex and both physical and other aspects of integrating the partial systems and sub-systems are challenging, the engineer designing the system is likely to require a very close understanding of both the partial systems/sub-systems and the manufacturing process. As the inter-relationship between sub-systems and the overall system is complex, similar or identical knowledge is likely to reside at many levels. Over time, industry may find it most efficient to consolidate to remove unnecessary duplication.

● However, in sectors where the relationship between platform and sub-systems/partial systems is less complex, and in particular where innovation is being driven primarily at the sub-systems/ partial systems level, it may be beneficial to keep the systems engineering tasks separate from the platform manufacturer, to ensure maximum openness to innovation from other suppliers. This might imply a design house construct, or an alliance (consortium, joint venture, or looser arrangement) where tasks, roles, risks and rewards are shared between members.

● It is generally desirable, and often essential, for the UK to have an on shore engineer of the overall system, particularly for supporting equipment in-service. However, even where the MOD has procured overwhelmingly from a UK-based supplier before, if the competitive environment is healthy, then unless other considerations prevent overseas supply[5], there is no reason in principle why another supplier could not establish a similar capability in the UK for new equipment. The existing base is likely in most cases though to give the current supplier in that sector a competitive advantage.

● Where a competition is kept open to overseas suppliers or a cooperative arrangement is pursued, unless the cost of retaining national flexibility is prohibitive or we are content to maintain the same configuration as another nation, MOD and industry must ensure that all necessary systems-level engineering knowledge to support and upgrade the equipment once in service is shared with an on shore systems engineer.

● Innovation once in-service, both for improving availability and enhancing other aspects of military capability, requires access to the overall systems-level knowledge, but may not occur uniquely in a platform manufacturer. This is particularly the case for systems where the platform is relatively less complex but the sub-systems and partial systems have decisive influence on the military capability sought. In such cases, it will be particularly important either that the contractual mechanisms allow value to flow through to the layers which are actually producing the innovation, or that sub-system/ partial system engineers are able to tender for upgrades directly.

Applying this analysis across the sectors

B1.42   This analysis demonstrates and explains the different conclusions across the individual sectors considered in this Strategy.

B1.43   Submarines are extremely complex, rely on sensitive technology, and critical aspects cannot be procured from offshore for security reasons. The sector is currently split into a number of monopoly suppliers (BAE Systems, DML, Rolls-Royce Marine) which contract direct with MOD, which is in turn a monopsony for the systems, sub-systems and partial systems provided, though total on-shore manufacture of every item is not required. Each of the main suppliers, and the submarine support business, is likely to contain some elements of the critical systems engineering knowledge. Opportunities exist for greater efficiency here, combined with action to establish a more consistent workflow. However, industry needs to reshape itself to protect both the systems engineering and manufacturing capability but in a more coherent and efficient form, with significant increases in productivity. The model needs to evolve, with industry, towards nomination of a single preferred systems engineering entity.

B1.44   We require a capability in UK industry to engineer complex surface ships at systems level, with enough familiarity with the manufacturing process to be able to fulfil that function. We have in recent years operated a system of competition in stages by project. However, again there is more capacity in the industry than will be required in a few years, and the systems engineering capability is likely to be duplicated and sub-optimised across several companies. Nor are potential synergies with support business being realised, despite largely the same companies being involved. The systems engineering capability needs, along with the rest of the industrial capability in this sector, to be refocused to maximise the relationship between in-service support and upgrade, and sized based on MOD's future needs and a realistic assessment of military export potential, if it is to maximise productivity.

B1.45   In the Armoured Fighting Vehicles sector, BAE Systems Land Systems has supplied 95% of the current inventory, and the associated systems engineering knowledge needs maintaining and developing; we will work further with the company to investigate mechanisms for ensuring this. However, the global market for future AFVs remains competitive, with much scope for innovation and new technology in FRES in particular. When these future systems are brought into service, a systems engineering capability will be required with the highest levels of systems engineering, skills, resources and capabilities based in the UK. The same features – high concentration of knowledge relating to the existing fleet, but a healthy international competitive environment – apply to the helicopter industry, where AgustaWestland's systems engineering capability needs sustainment in the short term and where partnership to drive improvements in support to the current fleet is natural, but competition remains an option for the medium term.

B1.46   In the general munitions sector, BAE Systems has the vast majority of the existing business, but there remain niche capabilities abroad which may meet future needs. We have therefore adopted partnerships with BAE Systems and other suppliers for support of the current inventory and are considering ways in which we can rationalise the through-life management of munitions, without ruling out the prospect of global competition for future projects at this stage.

---

[5]   e.g. the impact on sustaining the requisite level of knowledge in the supplier supporting the in-service equipment

Hercules C-130J under construction.

B1.47   For <u>C4ISTAR</u> there are a number of healthy companies with systems engineering skills in the UK, and given civil opportunities in this sector global competition by project seems likely to be sustainable for the foreseeable future. A similar picture applies for <u>CBRN</u>, where it is possible that partnering may offer some opportunities, but where competitive approaches are sustainable should these not demonstrate improved value.

B1.48   The challenge in <u>complex weapons</u> is whether the capability can be sustained. In principle, this is of high priority to us. But unless industry can restructure and deliver a viable proposition within the funds available, we may have to accept that we cannot sustain this capability onshore, recognising the implications for operational sovereignty that would entail. We have however decided that while we require this capability for in-service torpedoes, we do not require a sovereign capability on shore to systems engineer complete future torpedoes, though we do require the capability to engineer and integrate the algorithms and homing heads.

B1.49 In <u>aerospace</u>, we require onshore systems engineering capability to support through-life our existing and planned fast combat jets, although in the case of the Joint Strike Fighter this is limited to the ability to integrate and upgrade UK weapons onto the system as part of Team Lockheed (with our interest – on both cost and military effectiveness grounds – lying very clearly in preserving system capability commonality and coherence with the US). As such we regard it as essential to work with BAE Systems – the only company in the UK able to contribute at the top tier to international programmes in this sector – to sustain its systems engineering understanding of these platforms. But we will also attach importance to sustaining the systems engineering capabilities at the partial and sub-systems level – which in certain cases reside elsewhere in the supply chain – that will be required to provide for their maintenance and upgrade through-life.

Definition

B2.1   The Maritime Sector is that element of the Industrial Base which designs, builds, supports and disposes of all naval platforms and systems. It encompasses ships, submarines, and their integral systems; including propulsion, services, combat systems and combat system elements. It draws extensively on other sectors, such as Guided Weapons, Aerospace and C4ISTAR (Command, Control, Communication and Computers, Intelligence, Surveillance, Target Acqusition and Reconnaissance). Maritime capability is delivered by the effective integration of platforms and systems, and their through-life support.

Future CVF & JCA (Computer generated image).

Strategic overview

B2.2   The 2004 Defence White Paper, *'Delivering Security in a Changing World – Future Capabilities'* , emphasised the importance of versatile maritime expeditionary forces to project power across the globe in support of British interests and delivering effect on to land at a time and place of our choosing. Future maritime operations are likely to follow a similar, expeditionary pattern to those conducted recently. The sea offers an opportunity for UK Forces to operate with a degree of security and persistence, without reliance on the territory of others for basing. These factors, in particular the need for freedom to operate in an uncertain world, make the sea a very attractive location from which to project power. To take advantage of this the Royal Navy will in future need to be an agile, network enabled expeditionary force able to switch between missions and tasks and to interoperate with chosen allies. The force will have the ability to deliver and sustain a full range of missions: from small highly focussed interventions with Special Forces, to large, high intensity coalition operations, securing key influence in the process. This versatile maritime force will be capable of winning safe theatre entry for the deployment of joint forces. Through amphibious operations and a full range of medium scale offensive air effort, the versatile maritime force will deliver Maritime Strike and Littoral Manoeuvre to achieve decisive effect on the land.

Equipment Programme

B2.3   We are currently in the middle of a substantial modernisation programme that will enhance the capabilities of the RN. It has particular emphasis on fewer but more capable platforms, focusing on the capability to conduct expeditionary operations.

B2.4 The two planned **Future Carriers (CVF)** will be the biggest surface ships ever to be built in the UK - and will carry a strike package of Joint Combat Aircraft (JCA). The CVF programme is subject to an incremental approvals process: Target In-service Dates (ISD) for the two vessels will be agreed when the manufacture phase is approved. Given that both France and the UK are embarking on major, complex carrier procurement projects, we are examining areas of mutual benefit and opportunities to deliver economies. It is for industry to put forward proposals which will be judged on their merits and in light of national policies. It has been agreed with France that for co-operation to work, it must deliver cost savings and must do so without delaying UK or French programmes.

In 8 days the RN assembled off the coast of Africa a Joint Force of over 3000 RN, Royal Marines, Royal Fleet Auxiliary and RAF personnel, in support of the UN in Sierra Leone

B2.5   The **Type 45 Destroyer** will provide the RN's primary Anti Air Warfare capability for over thirty years. It is a versatile warship that will provide exceptional detection and air defence capability when the first of class, HMS DARING, enters service . This capability is centred on the Principal Anti Air Missile System (PAAMS), delivered through a collaborative consortium in EUROPAAMS; and SAMPSON, a UK Multi Function Radar under development with BAE Systems. Up to eight Type 45 Destroyers are planned to enter service in the next decade.

B2.6   A **Future Surface Combatant (FSC)** study is looking at how the capability currently provided by the Type 22 and Type 23 frigates might be met in the future. No decisions have been taken, but our current assumption for planning purposes is a two class platform solution. The **Future Mine Counter-Measures Capability** is also being examined.

B2.7   The **Astute Class** will be the most advanced and powerful attack submarines the Royal Navy has ever operated and will play a key part in our defences for decades to come. With improved communications, a greater capacity for joint operations and the ability to carry more weaponry, the Astute-class submarines will deliver a marked increase in the flexibility of our attack submarines. Three Astute Class nuclear powered submarines are on contract with BAE Systems and due in-service in 2009, 2010 and 2012, with potential for a further 5, subject to affordability.

B2.8   The future **Amphibious Capability** will be built around specialist shipping consisting of two **Landing Platform Docks (LPD)**, one **Landing Platform Helicopter (LPH)**, an Invincible Class aircraft carrier in the LPH role, and four **Landing Ship Dock(Auxiliary) (LSD(A))**. The LSD(A) class is expected to remain in-service for around 25 years. Additionally, CVF will be deployable in a secondary role as a Helicopter Carrier.

A Landing Craft, Air Cushion (LCAC) from 539 Squadron Royal Marines approaches the well dock of HMS ALBION.

B2.9   The **Military Afloat Reach and Sustainability (MARS)** programme is a significant planned investment in a new integrated approach to Afloat Support, combined with investment in life extensions for retained platforms. The MARS system-of-systems may include Fleet Tankers, Joint Sea Based Logistics and Fleet Solid Support vessels.

B2.10   **Type 23 Frigate Capability Upgrade Programme** is complementary to the FSC concept and potentially extends the life of the Type 23 Frigate. Capability upgrades are planned for the combat system, with updates to address structural strength and platform systems to follow.

B2.11   The **Trafalgar Class** SSNs (nuclear powered submarines) are nearing completion of a world-leading sonar and combat system improvement programme. This will ensure the submarines remain effective for the remaining life of the class.

B2.12   The **Vanguard Class** SSBN (nuclear powered ballistic missile submarine) main sonar inboard electronics are about to be delivered by a technically and commercially open systems solution, marking a pioneering and significant change in our approach to through-life capability sustainment.

Vanguard submarine.

B2.13   Capability investigations are underway, exploring the utility of **Minor War Vessels** for Maritime Interdiction Operations and an Anti-Fast Inshore Attack Craft capability.

B2.14   The **Offshore Patrol Vessel** replacement for the Falkland Islands Patrol role will be through a leasing arrangement with VT; its expected ISD is 2007.

B2.15   **Support** to warships, submarines and Royal Fleet Auxiliaries, including their update and upgrade, represents a significant element of a platform's whole-life cost; for example, for CVF the initial procurement will account for around one third of total through life costs. In recent years the total amount of support

work has diminished as a result of force level rationalisation, but the planned life extension of Surface Combatants moderates the reduction out to 2030. The level of future support still represents significant opportunities for UK industry.

Indicative planning assumptions

B2.16   The assumed spend profile in the maritime sector is expected to grow over the next ten years, providing a very strong programme of work for UK shipbuilding as T45, Astute, CVF and MARS work comes on line. This is followed by a longer term downturn as these major programmes come to an end. As a customer, we cannot afford and do not need to maintain the current pace of successive new platforms once the new ships are in service. This has implications for both new procurement and the volume of support business required. As the graph demonstrates, a very significant amount of resources - around half the amount the Department spends annually on the maritime sector - are consumed in supporting naval equipment.

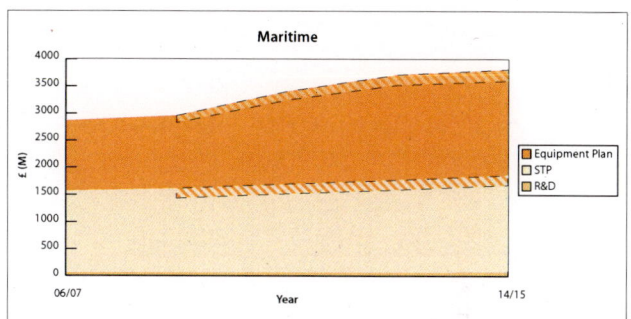

Figure B2(i) Illustrative spend profile.
The above graph shows indicative spending in this sector over the next ten years. The figures from 08/09 are illustrative and include a range in order to emphasise the potential for shifts in investment priorities after the end of the current Spending Review period. This is prudent planning which does not distort the overall illustrative picture of general trends.

What is required for retention in the UK industrial base?

B2.17   Retention of onshore capability is driven by two fundamental strategic requirements: the need to develop and support military capability throughout its life; and the ability to mount operations from the UK base. To meet these two requirements we have identified six strategic themes supported by a breakdown of specific capabilities. Where these are at a high level the maintenance of each capability is critically dependent upon retaining access to associated skills, facilities, processes and underlying technologies.

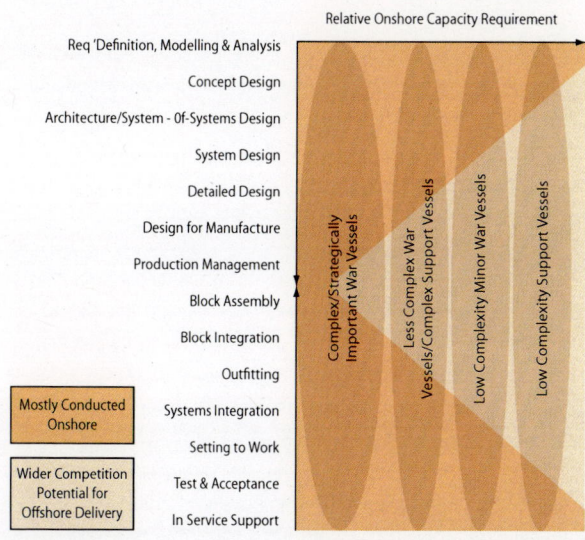

Figure B2(ii).

B2.18   Not all key capabilities must be exercised onshore for every project. The strategic need for onshore execution will be judged on a case by case basis (Figure B2(ii) illustrates this point); with the proviso that offshore delivery should not challenge the viability of key capabilities in the Maritime Sector as a whole. Using this model we can distinguish between that which must be executed onshore; and that which may be competed more widely, but might need to be executed onshore for reasons of sustainability or commercial viability.

Strategic capabilities for retention onshore:

Maritime systems engineering resourse: it is a high priority for the UK to retain the suite of capabilities required to design complex ships and submarines, from concept to point of build; and the complementary skills to manage the build, integration, assurance, test, acceptance, support and upgrade of maritime platforms through life.

Shipbuilding and integration: there is no absolute requirement to build all warships and Royal Fleet Auxiliary vessels onshore, but a minimum ability to build and integrate complex ships in the UK must be retained.

Submarines: for the foreseeable future the UK will retain all of those capabilities unique to submarines and their Nuclear Steam Raising Plant (NSRP), to enable their design, development, build, support, operation and decommissioning.

Maritime Combat Systems: the ability to develop complex maritime combat systems is a high priority for the UK, and their integration into warships and submarines is an essential onshore capability.

Maritime support: the UK shall retain the ability to maintain and support the effectiveness of the Fleet, including incremental acquisition, generating force elements at readiness, and meeting urgent operational requirements.

Maritime systems and technologies: it is a high priority to retain onshore research, development and integration of specific key maritime systems and technologies.

Maritime systems engineering resource

B2.19   The systems engineering resource includes: design expertise from early concept through to design for manufacture; all elements of maritime project management and the ability to specify and manage complex warship integration, test & acceptance at the platform and system-of-systems levels. These skills are as relevant to the through-life management of military vessels as they are to the front end procurement process.

B2.20   Maintaining control of the procurement and support processes as an intelligent customer is essential, regardless of where they occur. During initial procurement and throughout service, we must be able to manage the product risk associated with complex maritime platforms, particularly for the first of a new class of vessel. We are also required to fulfil our duty as a safe and competent owner and operator of our assets; and we will regularly use industry to provide supporting advice. Therefore, retention of the Maritime Systems Engineering Resource must encompass the expertise necessary to generate and support military capability throughout the acquisition lifecycle.

Through-life capability management. A good example of this in practise is the refit of HMS ILLUSTRIOUS to prepare it for a new dedicated strike carrier role. It is also a good example of how the UK shipbuilding industry can rise to such challenges. HMS ILLUSTRIOUS was a 30 month, £120M refit, to deliver an extensive upgrade package within an ambitious timescale. It came in under budget, enabling the savings to be re-invested in additional upgrades to the ship during the refit. Central to this success was a triangular partnership between the contractors, the MOD and the ship's company. The NAO cites this as a good practice example in its recent report - Driving the Successful Delivery of Major Defence Projects.

Shipbuilding and physical integration

B2.21   In a change to the previously stated Defence Industrial Policy (DIP), there is no absolute sovereign requirement to construct all our warship hulls onshore. We have revised our approach which concentrated solely on hull construction, now to consider sovereignty of the high-value capabilities needed for our operational independence.

B2.22   We need to build onshore to the extent that it sustains the ability to design and physically integrate complex warships. Furthermore, since warships are rarely prototyped, we need to ensure that we retain the ability to learn and adjust designs whilst the first of class is being built. Steel may be cut when the design is relatively incomplete compared to other military platforms; feedback during the production process is critical to ensuring that the platform meets the requirement as intended.

Type 45 Destroyer.

B2.23   The build of warships extends beyond the simplistic view of steelwork and its assembly, incorporating an amalgamation of skills, facilities, technologies and knowledge. In particular, it is the high complexity, value added aspects of ship build and platform integration that must be maintained under UK sovereignty: this includes specialist hull construction involving signature amelioration, Nuclear Biological Chemical Damage Control requirements, and complex fabrication and assembly technologies. These capabilities can be maintained in the long term only by their continued employment in suitably representative programmes of work.

B2.24   There is no requirement for fabrication of basic structures in the UK per se; however, mounting military operations from the UK base (including the fit of specific equipment for the operation in question), requires the relevant facilities and skills to be available onshore. Additionally, it is not effective to develop from scratch the most advanced, high-value skills needed for specialist hull construction or complex assembly tasks. There must be sufficient fabrication onshore to sustain a skills development path for workers to learn their trade and progress towards the most challenging tasks.

B2.25   When determining where aspects of a programme should be executed, straightforward cost considerations cannot be taken in isolation. We must also consider the strategic requirement for an industrial programme, sufficient in volume and complexity to deliver higher-end capabilities. Programmes that will tend towards total onshore delivery are those where the complexity (typically 'packing density' or outfit to steel work ratio) is high: the management and overhead of an offshore fabrication effort becomes less attractive when the high value aspects of a programme significantly outweigh the low order fabrication costs. This is especially true when a high level of outfitting is conducted at the same time as block construction.

The ratio of Combat System to platform costs is typically 2:1 for complex vessels; for Type 45 it is in the region of 60% for the Combat System against 20% hull costs.

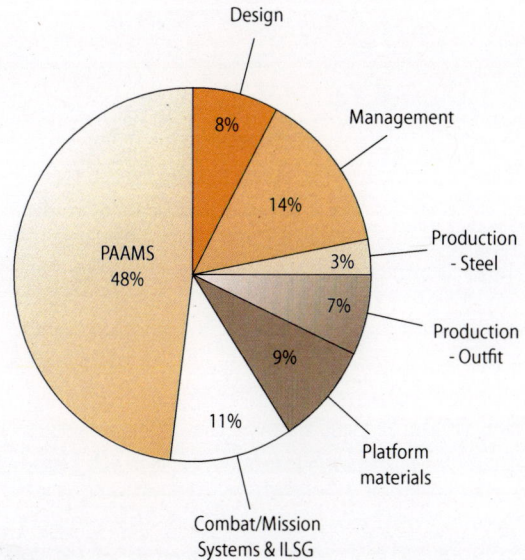

Naval Vessel Costs - Indicative split
Source: BAE SYSTEMS (T45 Whole Warship)

- Design 8%
- Management 14%
- Production - Steel 3%
- Production - Outfit 7%
- Platform materials 9%
- Combat/Mission Systems & ILSG 11%
- PAAMS 48%

Figure B2(iii).

Submarines

B2.26   The UK's fleet of nuclear powered submarines requires a specialist subset of skills within the maritime industry. We have duties of nuclear ownership and commitments to the USA which can only be fulfilled by close control of an onshore submarine business. Therefore, it is essential that the UK retains the capability safely to deliver, operate and maintain these platforms, without significant reliance on unpredictable offshore expertise. This delivery spans from conceptual design through to disposal, and includes the management of submarine and nuclear safety; all underpinned by appropriate science and technology. Some submarine sub-system elements may be sourced from abroad, but only under appropriate arrangements that guarantee supply, or from a sufficiently broad supplier base to assure access and availability.

B2.27   Deep scientific and technical advice on hydrodynamics, manoeuvring & control, propulsor technology and atmosphere control are specific capabilities essential to submarine performance. Structural acoustic engineering design is not readily available from the broader marketplace and has to be maintained within the specialist submarine industry. Submarine hull and infrastructure design and construction require the use of specialist techniques, for example particular welding and fabrication processes. These specialist underpinning key capabilities must be sustained in the UK.

B2.28   The ability to manage Nuclear Steam Raising Plant throughout its life-cycle, including the fuel elements, is a strategic capability that must be retained onshore. This includes design and development, manufacture, test and evaluation and decommissioning. An irreducible minimum level of associated facilities, intellectual resource and supporting technologies must be provided within the UK or under arrangements that guarantee UK control and safe ownership.

Astute (Computer Generated Image).

Maritime Combat Systems

B2.29   A Combat System is a sophisticated and complex system, ongoing development is essential if interoperability and military advantage are to be maintained. Combat System engineering consists of two complementary endeavours: the logical development of sub-systems into a single Combat System; and the physical integration of the Combat System into the platform, to deliver the platform's military capability. These two aspects of Combat System engineering apply equally for both surface ships and submarines.

B2.30   Not all elements of a Combat System must be developed and provisioned onshore; but it is strategically important to be capable of developing a single integrated Combat System. Maintaining control of specification, design, integration and acceptance is fundamental to initial procurement and through-life management of the Combat System, including spiral development and incremental acquisition. This dictates absolute involvement at the front edge of procurement and an ongoing relationship with a sovereign Combat System Design Authority.

The Type 42 Class of Destroyers has undergone a major architectural redesign and five further capability upgrades in the last 12 years.

B2.31   Physical integration of a Combat System into a maritime platform requires co-operation between the systems engineering organisation that maintains the design architecture of the platform and the Combat System design authority; given the likelihood of ongoing change through-life, this needs to be an enduring relationship. This high value-added aspect of shipbuilding must be retained within the UK maritime industrial base, if through-life development is to be pursued for complex or strategically important platforms.

Maritime support

B2.32   Support of the UK Fleet has traditionally been divided between Operational Support and Refitting, each with very different requirements and characteristics. However, the division is becoming increasingly blurred by an approach to routine upgrade known as 'Fleet Time Fitting', which is undertaken during periods in harbour for vessels at higher states of readiness. Onshore ability to conduct both Operational Support and Refit is strategically essential, but largely for different reasons and at differing scales.

B2.33   The need for Operational Support is equally applicable to warships, submarines and RFAs. Implicit in Operational Support is the ability to mount operations from the UK base through rapid force generation; it involves bringing units to increased levels of readiness, including the installation of mission specific equipment, and the provision and integration of equipment to meet urgent operational requirements. These tasks frequently require a high speed cycle through the acquisition process, and involve classified military capabilities and the handling of highly sensitive material. Therefore, key discriminators for provision of Operational Support include maintenance of national security and assured access to meet operational planning assumptions. Conduct of system upgrades by 'Fleet Time Fitting' increases the overall operational availability of the Fleet, but introduces similar demands to those of rapid force generation, albeit in slightly less demanding timescales.

In preparation for Operation TELIC, more than 30 warships, submarines and RFAs were fitted with over 120 operational enhancements in less than one month.

B2.34   The infrastructure required to conduct refits is extensive and not readily regenerated once lost. A level of surface ship refit capability must be retained in the UK to ensure guaranteed access when required, including for urgent operational support. An onshore refit capability becomes essential when security needs safeguarding, force protection is

a significant issue, or control of the programme is strategically necessary. Contingent docking and recovery from operations will require a UK dockyard, especially as embarked ammunition is often involved. For the less complex platforms, refits may be conducted offshore (e.g. RFAs and some minor war vessels) once sensitive equipment has been removed or security concerns, including force protection, otherwise safeguarded. The requirement to refit the submarine flotilla onshore is absolute.

T23 Frigate.

Maritime systems and technologies

B2.35   Running through each of the strategic themes is the need to sustain sufficient research and technology investigation to develop and maintain maritime domain expertise. This supports the UK in remaining an intelligent customer, even when buying elements from offshore, and is particularly pertinent to matching capability to threat. In the past, we have held sufficient research capability in-house, but it is increasingly developed and sustained by industry.

UK Mine Countermeasures and Uninhabited Underwater Vehicles expertise enabled evaluation and adaptation of a US commercial reconnaissance vehicle, which is now in service with the RN.

B2.36   The UK has a strategic advantage in many key platform and Combat System technologies and systems. These military capabilities are often in sensitive areas and have high security classifications. For the purposes of operational and strategic security, or assured access at times of tension or conflict, onshore retention of key research and development is a high priority. Onshore expertise also enables the exploitation of wider research to deliver systems that meet UK capability requirements. Retention of these key capabilities is fundamental to maintaining the battle winning edge.

Overview of the current maritime Global and UK Market

Global overview

B2.37   Worldwide **commercial** shipbuilding is mainly in Asia (Korea, Japan and China), which has around 70% of world production. With about 20% of world ship production, Europe is competitive for the more complex platforms such as passenger carriers and specialist vessels.

B2.38   Global **military** shipbuilding is dominated by the USA and Europe. In the US, ownership has consolidated into two main shipbuilding companies and two companies providing major sub-systems. Europe has twelve major military shipbuilding companies, with the bulk of these in UK, France, Germany, Spain, Italy, and the Netherlands: having consolidated from a larger industrial base further rationalisation seems likely. Similarly, there are extensive military ship repair facilities throughout Europe and within the US, many still controlled by national governments; consolidation and rationalisation is also evident in this area. To date, rationalisation has not extended across borders, although some cooperative programmes have been pursued by European governments. Retaining national military support facilities is widely seen as an essential requirement for mounting and supporting operations of a first class Navy.

The UK sector

B2.39   The contraction of the UK shipbuilding industry has been driven by fierce competition for commercial shipbuilding work, primarily from within Europe and the Far East. The UK industry is no longer sufficiently competitive to win substantial amounts of traditional merchant shipbuilding, especially where extensive conventional steelwork is involved. However, the industry remains internationally competitive on high-value conversion and refit work, and on specialist builds such as luxury yachts.

B2.40   A reduction in UK warship building has mirrored the parallel reduction in the number of platforms required by the Royal Navy. Nevertheless, the UK remains a major provider of warships, ranked in the world's top four alongside USA, Germany and France. MOD is the UK shipbuilding industry's biggest customer, and naval ships comprise around 85% of those being constructed in UK shipyards. We will spend several billion pounds in the next decade to procure new ships and submarines. The potential for exports to help sustain the UK industrial capability should not be underestimated. The RN is a valuable asset to industry in promoting export business. However, UK new builds for export are a small fraction of the domestic output, whereas European states export a significant proportion of their total build. This reflects the global demand for modestly priced frigates, rather than the high-end complexity currently represented by the majority of UK shipbuilders' portfolios.

France and Germany together have more than 60% of the military export market; Germany producing twice as much for export as for domestic use.[1]

B2.41   The maritime support workload has also reduced in recent years, both as a result of force level reductions and new rationalized maintenance techniques. Whilst some increase in demand for updates and upgrades will moderate this trend, the UK exhibits over-capacity in support facilities. Existing suppliers have not been incentivised to rationalise, as keenly competitive bidding has driven down prices, limiting funds available for the short-term investment required. The repair yards have therefore experienced fluctuating work loads.

B2.42   Ownership of UK warship yards has consolidated to two main companies with the skills necessary to design, manufacture and integrate complex warships: BAE Systems (Naval Ships and Submarines) and VT Shipbuilding; with further capacity at Swan Hunter. DML and Babcock Engineering Services have design capability and fabrication skills but, together with FSL, essentially deliver surface ship and submarine support (including upkeep).

B2.43   Areas of critical expertise such as design and systems integration skills exist throughout the industrial base, not simply within the manufacturing sector. For example, BMT, QinetiQ and Three Quays have expertise in naval design and systems engineering; QinetiQ having the additional capacity to undertake research. Other large companies without shipyard infrastructure contribute significant capabilities. For example, Rolls-Royce Marine design and manufacture submarine nuclear propulsion and marine gas turbines; Thales Naval is a leading Combat System design, engineering and integration company, whilst supplying specific systems such as sonar; Ultra is proficient in underwater systems and naval Command and Control. More than half the unit cost of a naval vessel lies with firms other than the shipbuilder, and we recognise the importance of small and medium enterprises as part of this mix, whether within the supply chain of primes or those that work directly with the MOD. Many of the higher order capabilities are dependent on the specialist skills and expertise of SMEs. SMEs' ability to meet our requirements is an important consideration.

Application of commercial capacity to defence

B2.44   There are clear differences between warship and commercial shipbuilding: the cost of a warship is typically 70% systems, 30% hull construction and outfitting; by contrast, for a commercial ship the figures are typically 20% systems, 80% hull construction. The underlying skill sets and processes for warship work are not available in yards focussed on the commercial sector. In general terms, the more war-like the vessel, the more complex the ship: this does not necessarily apply to hull fabrication, but does apply to many aspects of design, outfitting, military system integration, test and commissioning. Naval shipbuilding is specialist work and demands significant assurance regimes, engineering and professional support, whose underlying skills take time to build and effort to sustain.

B2.45   The differences between military and commercial shipbuilding need not necessarily exclude commercial shipyards from military shipbuilding. Their expertise potentially is relevant to less complex auxiliary and support vessels, where commercial design and production techniques offer considerable efficiencies over warship construction practices. The wider commercial sector also offers a benchmark against which military yards can set performance improvement targets, taking into account the increased complexity of military shipbuilding. Non-warship facilities also undertake a valuable supporting role in fabrication and other work, particularly during periods of peak demand for facilities and resources. The wider industrial base has system integration experience, but this is not directly comparable to the complexity of warship integration. Nevertheless, there are some useful lessons to be learned from the Alliance/partnering approach the wider industry adopts, the potential of which will be exploited by the CVF programme.

---

[1] 'Military and Commercial Shipbuilding' RAND (2005)

UK Military Shipbuilding Skills Base

- UK military shipbuilding requires a highly skilled work force that can be confident in an enduring and stable career path. This is particularly true of the high value skills, knowledge and expertise demanded for the delivery of complex warships.
- The ratio of white to blue-collar workers in commercial yards is 1:6, in military yards it is about 1:1·7.
- In some areas, industry is confident of its ability to generate capability rapidly should the need arise, steelwork fabrication being a key example. However, many military standards (such as for welding and surface flatness) are higher than for commercial work.
- Research suggests that when shipyards lay-off workers, 70% of them leave the industry and are unavailable for re-hire by their former employer[2].
- There is a perceived skills shortage in specific capability areas. For example, industry agrees that design engineers are in short supply; and the intellectual support of underpinning science and technology is also fragile in some areas.
- Demographics are likely to feature as an increasing challenge in the sustainability of this workforce and the delivery of the Maritime Sector's key capabilities.

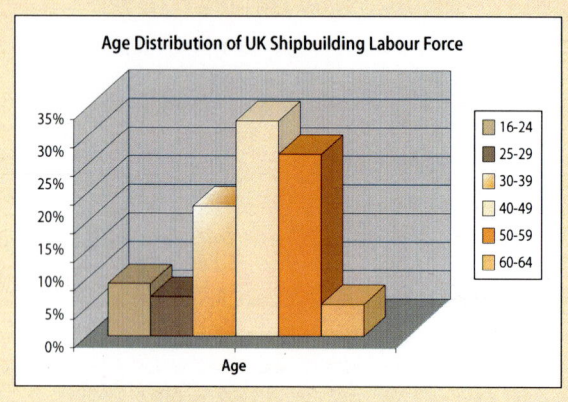

Age Distribution of UK Shipbuilding Labour Force

Source: 'Outsourcing and Outfitting Practices'. RAND 2005

Sustainment strategy

To maintain the key capabilities, a vibrant onshore forward programme is required, focusing on high value activities.

B2.46   The planned maritime forward programme represents a healthy customer order book for the industry and is likely to sustain UK employment in the maritime sector well into the next decade. The UK Maritime Industrial Base currently possesses the key capabilities required to support this programme. Furthermore, the UK has the industrial capability to design, manufacture and support all UK Fleet surface ships, submarines and auxiliaries, but may not have the fabrication capacity to absorb the full programme at its peak. However, the high volume of programmed shipbuilding activity cannot be sustained indefinitely. Beyond the peak of activity for CVF the potential work available to UK industry reduces to a steadier state by around 2016. The future for UK shipbuilders lies in high value design, systems and sub-system assembly and integration; plus specialist and novel hull construction capability, particularly where there is a high outfit to steel ratio, as exhibited in complex warships.

[2]   *'Reducing the strains in the labour force available for warship building in the UK'. Furness Enterprises Ltd. July 2003.*

The UK's Maritime Industrial Base must deliver improvements in its performance.

B2.47   To deliver an affordable forward programme the maritime sector faces considerable challenges, including industry's ability to control costs. The UK maritime business is characterised by high and increasing overheads, and has a skills base spread across too many entities. Procurement strategies and commercial arrangements have not adequately incentivised or enabled rationalisation and efficiency improvements. The sector has failed consistently to deliver satisfactory performance, with several high-profile maritime projects encountering delays and cost increases. The business must be streamlined for greater efficiency and profitability, whilst mirroring UK demand and maximizing the opportunity for export. The UK will need to buy warships and submarines for the foreseeable future, but the clear trend is for fewer, more capable platforms, with longer operational lives and increased opportunity for regular upgrades in response to new technologies and threats. The ability to do so will depend upon us working together with industry to address the fundamental issues of affordability and productivity.

Challenges for UK Shipbuilding

Independent study has shown:

- Major UK Defence Acquisitions are typically behind schedule.
- Commercial ships are typically produced on time.
- Ship builders employ no consistent forecasting methodology.
- We must work with industry to better manage late changes.
- Late delivery of commercial ships attracts more punitive financial penalties than for military vessels.
- The commercial and military markets differ significantly in ship size & complexity, acquisition process, design and construction, and the work force skill sets and make-up.
- Industry restructuring and changed industry/ MOD processes could benefit the UK military programme and increase export opportunities.

Source: 'Monitoring the progress of shipbuilding programmes'. RAND 2005

Without improvements in performance, delivery of the forward equipment programme is threatened. Industry restructuring is a priority.

B2.48   The current situation is unsustainable and places huge pressure on the future programme. Whilst applicable to surface ships it is compounded many times over in the submarine domain, due to the high cost of entry for these specialist capabilities and the very high overheads for their continued delivery. Industry restructuring and consolidation is likely to be a key feature of any improvement programme, and fundamental to creating a viable and sustainable business to meet anticipated steady-state demand.

B2.49   In addition to horizontal consolidation the potential for integration of procurement and support delivery must be realised if efficiencies are to be generated. This offers the prospect of better management of through-life military capability, from delivery to disposal. It would also entail rationalisation of facilities and the skill base, delivering a more enduring and stable career path.

B2.50   In light of the serious financial challenges facing the industry, it is our view that consolidation should occur as a matter of urgency. This is particularly pertinent to the Submarine domain, but applies across the board.

The nature of restructuring is for industry to consider, but must be customer focused

B2.51   We will not micromanage industry's restructuring but it must be customer focused and we are likely to express preferences as different approaches emerge. We must be confident that consolidation will be beneficial to MOD and industry. We are considering potential models as they arise and these might involve some form of Government stake in how the industry develops. We also recognise that as the predominant client we are critical to improving the efficiency of the supply demand relationship.

We will pursue procurement strategies and commercial arrangements that are optimised for the sector to deliver three key objectives: a sustainable enterprise, better performance for MOD, and opportunities for attractive rates of return for industry.

B2.52   We will seek to employ more sophisticated strategies and arrangements that will be optimised for the sector. Competition will continue to be used when appropriate, especially for embedded electronics and marine equipment, but alternative approaches will be developed where they are necessary to deliver greater value for money and long term sustainability. As an example of an optimised approach the Future Carrier (CVF) project is being pursued through the CVF Alliance. This type of arrangement is well established in the oil and gas industries, but innovative for UK defence acquisition. It draws on the strengths, resources and expertise of all parties with rewards geared to the overall project outcome rather than maximising benefits to one participant.

Type 23 HMS SUTHERLAND.

There will be a minimum level of activity, or Core Work Load, necessary to sustain the key capabilities.

B2.53   We recognise that simply maintaining a minimum sovereign industrial base is not likely to be attractive to industry or to represent good value for money. To make the industry viable will require a through-life capability approach based on cost of ownership. Working with industry we will define a Core Work Load that not only would sustain the key capabilities, but also offer value for money and be commercially viable, allowing industry to scale its core capacity accordingly.

B2.54   The Core Work Load will contain all activity unique to submarines. For surface ships it is possible that only a proportion of the total programme in any given period may be required to sustain key capabilities. This core is likely to be centred on, though not necessarily restricted to, an onshore build capability for large complex warships. This activity will provide the necessary experience for the management of build, integration and testing across the wider maritime programme. The Core Work Load will include support activities required to prepare and deploy UK forces.

We will provide industry with visibility of a sustained demand to deliver this Core Work Load.

B2.55   We will seek to sustain this workload to ensure the retention of key capabilities and the viability of the business that delivers them. This will be achieved by viewing the forward programme as a set of projects that may be phased to balance required military capability, affordability and industrial sustainability. Clearly, flexibility will continue to be required as circumstances can change; but given the importance of sustaining a critical mass of onshore expertise, for both maintaining sovereignty and delivering value for money, sustainability impacts will be given serious attention when adjustments to the programme are being considered.

B2.56   The concept of project frequency, or 'drumbeat', is a response to this theme. For submarines we have endorsed, but not yet committed funding for a 24 month SSN build drumbeat. This scales the build capacity to be satisfied by the industry supply chain after the third Astute Class submarine (HMS ARTFUL); and sets the rhythm for the rest of the programme, notably support. The longer term surface ship production drumbeat is of the order of one new platform every one to two years, given anticipated force levels and platform life cycles. The concept of drumbeat is not restricted to major platform delivery, but includes discrete key capabilities, such as Combat System development.

B2.57   The Support work-rate is set by the size of the Fleet and the maintenance cycle, which is dominated by overhaul periods, and defueling for submarines. The new vessels (Astute Class, Type 45) will require less maintenance than legacy platforms. This combines with the reduced size of the Fleet to result in a lower and fluctuating maintenance demand. To counter this we are assessing alternative maintenance cycles with more frequent, less intrusive interventions, which will both smooth demand and improve readiness.

We will not pay a premium for capacity in excess of that required to deliver the Core Work Load.

B2.58   Projects within the maritime programme that exceed the Core Work Load requirement may be widely competed and potentially undertaken offshore if it does not prejudice the key capabilities. UK industry will be able to bid for this, capacity allowing. However, we will not expect industrial capacity over that required to meet the Core Work Load to have an adverse impact on the MOD's overall exposure to industry's overheads. When considering work outside the Core Work Load envelope, we will not make a simplistic distinction between entire platforms: the concept applies equally to discrete project elements.

B2.59   The CVF and Type 45 programmes represent a significant deviation from normal steady-state demand. It would be unwise to expand onshore capacity above current levels, only for it to contract rapidly after CVF delivery. Low complexity elements of CVF build are strong candidates for offshore provision, if UK steady-state capacity is exceeded and better value for money is offered elsewhere. After the Type 45 and CVF surge we will seek to ensure a managed transition to a more typical, less intensive build/integration activity. This will involve smoothing the work rate to sustain the Core Work Load.

Type 45 (Computer Generated Image).

We recognise the fragility of the design base and we will implement measures to exercise the capability when this is strategically necessary and can be shown to offer long-term value for money.

B2.60   Major design is a relatively infrequent activity naturally occurring just once per class. However, maintaining the platform design is a through-life activity, with updates and upgrades requiring significant design effort up until a platform's last refit (often with further application on disposal). By combining the new build and support design activities in a rationalised manner, a more sustainable capability is possible. This also offers the potential for whole-life cost reduction and capability enhancements, as well as long-term career paths for the associated engineers.

B2.61   CVF detailed design work will employ much of the nation's maritime engineering workforce to the end of the decade. However, early concept and architectural design requires a subset of this skilled workforce, which will need managed short term sustainment as their employment by CVF diminishes.

B2.62   Submarine design capability is at risk if long gaps emerge between first-of-class design efforts. The eleven year break between the design of Vanguard and Astute undoubtedly led to a loss of capability and impacted on the Astute programme. We now aspire to an eight year drumbeat to sustain the design capability through incremental improvements, both to drive down build costs and reduce subsequent support costs. In the short term key design effort will be focussed on improving these whole-life costs in the existing Astute design, particularly in areas that have direct benefit to subsequent classes.

B2.63   The submarine design programme will ensure options for a successor to the current Vanguard class deterrent are kept open in advance of eventual decisions, likely to be necessary in this Parliament. Cost-effectiveness will clearly be a key factor in any consideration of potential options, both submarine based and non-submarine based. For submarine-based options it will be very important that MOD and industry are able to demonstrate an ability to drive down and control the costs of nuclear submarine programmes. Industry will be fully engaged in ensuring that design efforts achieve the maximum impact in control of submarine build and support costs, so sustaining the potential for this significant future business and military capability.

Combat Systems sustainability and ongoing development will be promoted by the use of modern design and integration techniques, whilst facilitating integration of products from both large scale traditional suppliers and smaller enterprises.

B2.64   Combat System design and integration capabilities are a clear strategic imperative to deliver the required installed performance in maritime combatants. The adoption of planned and future upgrades will help to maintain the necessary suite of capabilities. In parallel, submarine and warship initiatives to converge towards a reduced set of core Combat System solutions will support the incremental approach. These common core Combat Systems will seek to exploit Modular Open System Architecture design philosophies, to enable continuous obsolescence management and affordable capability insertion across the Fleet.

Type 23 Frigate's Operations Room.

B2.65   The Surface Ship Combat Management System Convergence and submarine Common Core Combat System initiatives are both seeking to promote these strategies in the medium term. These initiatives have the potential to consolidate and retain the strategic capabilities necessary to form Combat System Architecture Authorities and support the specialist capabilities necessary to integrate modern high-technology sub-systems. A key objective is to exploit Open Architectures to allow SMEs, many from within UK industry and academia, to contribute niche capabilities in areas such as sensor algorithms, data fusion, security, and knowledge based systems.

B2.66   In the longer term we will investigate innovative methods of sustaining the UK's Combat System design, integration and acceptance expertise and associated facilities. We will welcome novel proposals from industry.

We will take specific measures to ensure sustainability of significant capabilities in 2nd and 3rd tier suppliers where these are at risk.

B2.67   We need further work to better understand the risks to 2nd and 3rd tier suppliers. Certain key capabilities have very limited sources of supply, which become fragile if they are not loaded or managed appropriately. Several levers exist to reduce exposure to this risk, ranging from increasing volume by amalgamating orders, to removing the critical component by redesign. We will work with primes to prevent the loss of key capabilities through failure of the supply chain. We are already moving in this direction with recent examples including procurement action to sustain the Astute Boat supply chain, and proposals to restructure aspects of the NSRP supply chain.

B2.68 Frequently a significant proportion of the escalation in project costs occurs through bought-in equipment. It is imperative for the MOD and industry 1st tier suppliers to ensure that they manage exposure to cost escalations throughout the supply chain.

We will seek to work together with industry to develop and sustain our own capabilities.

B2.69 It is essential that we sustain the qualities necessary for the MOD to fulfil its obligations as a safe and competent owner and operator of its vessels. In some specialist areas our capability is fragile. Action is now in hand to redevelop these areas and to actively career manage associated disciplines. We anticipate this will include working with industry, using secondment and joint working to develop knowledge for the benefit of both the MOD and the private sector.

B2.70 A range of measures are being applied to improve our performance and coherence. For instance, Director General (Nuclear), based in the DLO, has been appointed as the single focal point for delivery of nuclear submarine programmes across the MOD. We are committed to change that enables industry to perform effectively and address overall long-term sustainability. In particular, we are developing a stream of work known as the Maritime Industrial Strategy (MIS).

MIS will be at the heart of developing a sustainable relationship between the MOD and industry.

B2.71 We have been working with industry on the MIS for some time, looking at how we can best tackle these difficult sustainability issues. This work is concentrating on more clearly identifying the likely volume and timing of future business, and defining in greater detail how we plan to maintain the sovereign capabilities we require. This includes defining the Core Work Load in discussion with industry. In parallel, we expect industry to begin restructuring itself around the emerging Core Work Load. The success of the MIS is ultimately dependent on companies' willingness to work together and draw their own conclusions. However, we need improvements in quality and efficiency if our programme is to be affordable. The MIS needs to define the routemap to delivering this whilst sustaining our sovereign capabilities.

B2.72 MIS now embraces the Submarine Acquisition Modernisation (SAM) and Surface Ship Support (SSS) projects. These initiatives were launched to address growing concern at the performance of elements of the sector. By combining these projects, examining both procurement and long-term support improvements, we recognize that a viable and sustainable Maritime Sector is dependent on a more coherent approach across both domains.

We will move ahead quickly to begin making the most of immediate opportunities.

B2.73 Under the MIS, we will immediately start negotiations with the key companies that make up the submarine supply chain to achieve a programme level partnering agreement with a single industrial entity for the full life cycle of the submarine flotilla, while addressing key affordability issues. The objective is to achieve this agreement in time for the award of the contract for the fourth and subsequent Astute class submarines in early 2007. This will be matched by the implementation of a unified submarine programme management organisation within the MOD.

B2.74 For surface ship design and build, we aim within the next six months to arrive at a common understanding of the Core Work Load required to sustain the high-end design, systems engineering and combat systems integration skills that we have identified as being important. We expect industry to begin restructuring itself around the emerging analysis as set out above to improve its performance. We will build on the momentum generated by the

industrial arrangements being put together on the CVF programme to drive restructuring to meet both the CVF peak and the reduced post-CVF demand. For surface ship support, we will start immediate negotiations with industry with the aim of exploring alternative contracting arrangements and the way head for the next upkeep periods, which start in the autumn of 2006. Key maritime equipment industrial capabilities will be supported by the production of a sustainability strategy for these equipments by June 2006.

The high work load in the immediate Maritime Equipment Programme opens a window of opportunity for industry to do things differently.

B2.75 The increased demand of the next few years will diminish after the middle of the next decade. Although over-capacity offers the theoretical prospect of competition, this is unlikely to be sustainable in a shrinking market. Value for money may soon be delivered better through alternative strategies. For example, one fully loaded allocated stream of surface ship build might offer better value for money than several partially loaded streams in competition. We have been working to smooth out the long term cyclical demand for naval warships and provide a more predictable future for ourselves, and industry. But this more stable future can only be achieved if the design, manufacturing, support and integration capacity within the industry is matched to that pattern of demand. There is a clear need to streamline the businesses, making them more efficient and profitable, removing duplication and establishing clear centres of excellence, to meet our requirements and maximise the military export potential. This is good for the Royal Navy, the taxpayer and for the long term sustainability of the industry.

B2.76 Our shipbuilding industry needs to renew itself and there is a window of opportunity to do so, now. By taking this opportunity head on and tackling the challenges it presents, there can be a fundamental shift from seeking profit through volume, to profit derived from excellent delivery, long-term support, and the continual improvement of the military capability available to the front line.

HMS ARGYLL.

Armoured Fighting Vehicles

Definition

B3.1   Armoured fighting vehicles (AFVs) are bespoke land military vehicles optimised for close combat operations which possess appropriate levels of survivability, lethality and mobility to enable operations in a high threat environment. They perform in general utility as well as specialised roles and can be either wheeled or tracked.

Strategic overview

B3.2   The Defence White Paper of December 2003 stated that the UK requires a clear focus on projecting force, further afield and even more quickly than has previously been the case. This places an emphasis on speed of deployment, flexibility, and the agility required to deploy rapidly for a diverse range of expeditionary operations. We should maintain a credible warfighting capability to undertake demanding combat in all appropriate Military Tasks and at varying scales of effort. In sum, our combat power must be credible enough to coerce and deter effectively and, when called upon, allow us to disrupt and defeat an opponent.

B3.3   The Army is being restructured in accordance with the Future Army Structure (FAS) programme to provide a flexible and balanced land force structure consisting of a mix of heavy, medium and light capabilities. A key element is the medium force which, when grouped with joint assets, is termed the Medium Weight Capability. It will provide a responsive medium scale intervention force characterised by a high level of deployability (including elements by air), and greater levels of mobility, firepower and protection than are currently possessed by light forces. The Future Rapid Effect System (FRES) lies at the heart of this capability, but FRES is also required to replace obsolescent AFVs (e.g. Saxon, FV430 and Combat Vehicle Reconnaissance

(Tracked) (CVR(T)) that are increasingly costly to run and have declining relative capability which exposes our forces to operational risk.

B3.4   AFVs will therefore continue to lie at the heart of the Army's military capability for the foreseeable future as they enable operations across the spectrum of operations, from stabilisation to warfighting. The Army's two Armoured and three Mechanised brigades are the predominant users, along with 3 Royal Marines Commando Brigade with its Viking and Hippo vehicles. The current vehicle numbers are summarised below:

Fleet	Number
Challenger 2 (CR2)	385
CR2 Driver Training Tank	22
Challenger Armoured Repair and Recovery Vehicle (CRARRV)	81
Chieftain AVRE/AVLB/ARRV	119
Combat Engineer Tractor (CET)	73
Warrior	793
CVR(T)	1255
Shielder	30
FV430	1492
Saxon General War Role (GWR)	491
Saxon Patrol (Northern Ireland)	131
Fuchs	11
Viking	108
Hippo (Beach Armoured Recovery Vehicle)	4
Total	4,995

Armour for the leading Battle Group of 4 Armoured Brigade wait for future deployment at Krivolac, Macedonia Tuesday March 2, 1999. The British troops and equipment were being prepared as a contingency for the developing stuation in Kosovo.

Equipment programmes

B3.5 **Saxon** entered service in 1984 and is currently planned to reach its out of service date (OSD) in 2014. Other than Urgent Operational Requirements (UORs) to vehicles deployed to Iraq and Afghanistan, Saxon is not currently expected to undergo either a life extension or an upgrade programme.

B3.6 **Combat Vehicle Reconnaissance (Tracked) (CVR(T))** entered service (ISD) in 1972, and has had, and continues to have, an extensive series of life extension programmes (LEPs) and upgrades since the late 1990s covering conversion to diesel engines; equipping Scimitar variants with thermal imagers and navigation and target location systems; the ongoing Bowman communication conversion; and the fitting of Platform Battlefield Information System Application (P-BISA). The Swingfire anti-tank guided weapon, fitted to the Striker variant, is planned to reach its OSD in 2008. Thereafter, the Formation Recce overwatch capability will be met by the dismounted Javelin anti-tank guided weapon capability operating out of the Spartan variants until FRES enters service. The current OSD for CVR(T) is assumed to be 2014, although this is subject to ongoing review.

B3.7 The utility vehicle **FV430** entered service in 1965 and has had successive extensions to its OSD. It is increasingly difficult to support. In 2006 the Defence Logistics Organisation (DLO) will start to convert the powertrain for 500 vehicles in order to extend their life to the current OSD of 2015; this may be reviewed further and as such the MOD may need to convert a further tranche of vehicles.

B3.8 The **Warrior** family entered service in 1987 and has been successful on a wide variety of operations since entering service. To ensure it remains capable to its OSD it requires a Capability Sustainment Programme (CSP) which is now in its initial planning stages. On current plans it will improve Warrior IFVs' lethality, protection, ergonomics and availability probably through a new turret, cannon and Health and Usage Monitoring System (HUMS), as well as providing linkages to the Future Integrated Soldier Technology (FIST) project. The CSP will include the current **Armoured Battlefield Support Vehicle (ABSV)**, also in its concept phase. We plan ABSV to modify the remaining Warrior IFVs not required under FAS to provide an armoured support vehicle for armoured infantry and engineer units. Warrior's OSD has recently been extended to the early 2030s.

A Challenger 2 tank from the Queens Royal Hussars pauses during a patrol to observe a Serbian checkpoint a few hundred yards over the border from Kosovo.

B3.9 **Challenger 2 (CR2)** entered service in 1998 and has had a successful operational life to date. Currently, CR2 is being converted to Bowman and P-BISA. CR2 is not expected to leave service until the mid 2030s. We will therefore need to consider the requirement to conduct a CSP in around ten to fifteen years time to maintain its relative performance in terms of lethality, mobility and survivability.

B3.10 **The Engineer Tank System programme** will enter service in December next year and is currently in its manufacture phase. It consists of TITAN (an armoured bridge layer) and TROJAN (an obstacle breacher); these will replace the current Chieftain-based Armoured Vehicle Launched Bridge and Armoured Vehicle Royal Engineers. These vehicles provide the mobility, counter-mobility and survivability support to armoured and mechanised battlegroups. BAE Systems Land Systems is contracted to deliver 33 of each type plus training and support.

Titan.

Trojan.

B3.11 **PANTHER** is in its Demonstration Phase and is assumed to enter service in the second half of 2007. PANTHER provides Command and Liaison Vehicles to replace some CVR(T) (Spartan and Sultan), FV430, Saxon and Land Rover Truck Utility Medium for a variety of combat, combat support, combat service support and command support units.

Panther.

B3.12 **TERRIER**, currently in its demonstration phase, is due to enter service in December 2008 and will replace the Combat Engineer Tractor. A high utility combat engineer vehicle, there is potential to develop TERRIER into a wider family of engineer vehicles. BAE Systems Land Systems is contracted to deliver 65 vehicles and initial contractor logistic support.

B3.13 **Future Rapid Effect System (FRES)** is the Army's highest priority programme and will be the central pillar of a capable, coherent and highly deployable medium force. It plans to deliver a family of network-enabled medium weight armoured vehicles covering a wide range of combat, combat support and combat service support roles. It has an ISD planning assumption for initial variants in the early years of the next decade, with further tranches of vehicles providing incremental enhancements to capability thereafter. Production is currently expected to continue into the late 2020s in order to deliver the large number of vehicles required by the Field Army.

B3.14 The programme is currently in its initial Assessment Phase (iAP), in which the MOD is being supported by an independent Systems House, Atkins. The broad aims of the iAP are: to further define the capability required; develop affordable options to meet the requirement; develop optimum procurement and support strategies and to manage technology and supplier risk to acceptable levels. A number of competitively let Technology Demonstrator Programmes (TDPs) are also being run in order to de-risk some of the emerging technology that is likely to be used on the platform. Central to this are efforts to define and construct the electronic architecture, with open standards where appropriate, that will allow FRES to take its envisaged central role within the NEC. FRES will have a pre-planned product improvement strategy to grow the capability over time, as well as a strategy for maintaining a consistent modification state across the fleet during its long production run.

Indicative planning assumptions

B3.15 The key dynamic from the graph below is an assumption of an increasing Equipment Plan (EP) spend which represents the beginning of our major investment in the FRES programme. However, there is a hiatus in new platform development as the TERRIER, TITAN and TROJAN programmes have all now moved into production and envisaged major upgrades to the existing fleet are not planned until the latter half of this decade or the early half of the next. We currently assume a consistent long-term spend within the STP of approximately £200M per annum. We expect to spend approximately £140M in AFV research over the 10 year period, most of which will be absorbed in the FRES programme, included within the EP line.

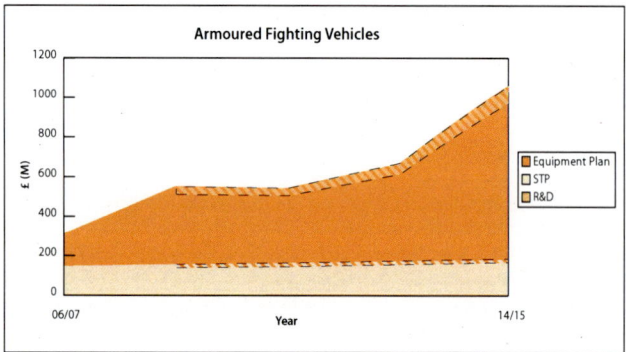

Figure B3(i) - Illustrative spend profile.
The above graph shows indicative spending in this sector over the next ten years. The figures from 08/09 are illustrative and include a range in order to emphasise the potential for shifts in investment priorities after the end of the current Spending Review period. This is prudent planning which does not distort the overall illustrative picture of general trends.

What is required for retention in the UK industrial base?

B3.16 The AFV fleet lies at the heart of Land Forces military effectiveness, without which the Land Force will be incapable of deployment or operations. Recent operations have shown their continuing importance and the significant demand for the capabilities they provide. The following table shows the number of AFVs that have been deployed in recent operations:

	Main Battle Tank	Heavy Support Variant	Warrior	CVR(T)	FV 430	Saxon	Other	Total
Falklands 1982				10				10
Gulf War 1991	221	124	331	584	861		357	2478
Bosnia (UN) 1995			52	10	47	65		211
Bosnia (NATO) 1996	29	26	98	176	235	79	10	653
Kosovo 1999	30	12	65	35	51			193
Iraq 2003	116	50	236	78	251		82	813

Figure B3(ii).

B3.17 There are compelling advantages to retaining a UK industrial AFV capability at a level which enables the UK to preserve the expertise it requires to maintain and upgrade the capability of current and future equipment, both in peacetime and for operational requirements. To make this possible, we need guaranteed access to the existing AFV fleet Design Authority (DA), BAE Systems Land Systems. But we will not do so at any cost. Indeed, the UK is willing to consider new AFV prime contractors being from offshore, so that we can access innovative capabilities from foreign sources which may not be present in the UK industrial base. This may also include aspects of the DA role, although given that our priority is to be able to upgrade, meet UORs and support existing fleets on-shore it will be necessary for the DA for new AFVs to establish or maintain some form of substantive and empowered UK presence (which could be through an onshore partner with access to the necessary IPR). With this in mind, the UK AFV industrial base needs to retain the following:

- **AFV Systems Engineering, Domain and Design Knowledge** - The UK's AFV industry needs to be capable of understanding the military context of the UK user. Such domain knowledge should include a developed understanding of the operational context and the Defence Lines of Development. It also includes the ability to understand as an intelligent customer how to design, develop and build a new AFV and the ability to integrate the platform **into the wider suite of capabilities that make up the network**. This knowledge also supports the skills required for the **through-life capability management** of AFVs, such as being able to understand how to undertake modifications, upgrades and complete UORs. Guaranteed access to design knowledge of the existing AFV fleet will also be important to underwrite the safety, legality and performance of these AFVs;

- the intellectual ability to **design, validate and interpret the results of AFV testing** should also be retained onshore, in order that the customer has complete assurance of the safety and capability of AFVs. Test and evaluation facilities do not necessarily have to be on-shore, except for sensitive sub-system testing;

- **critical AFV sub-systems** – Within AFVs are particular sub-systems which provide either battle-winning capability or strategic leverage with coalition partners. It is therefore critical that the UK industrial base retains the necessary critical mass to design, build and **integrate such sub-systems onto the platform** to ensure continued guaranteed access. These technologies include:

    - Integrated survivability solutions, notably special and electric armour;
    - Electronic architecture, particularly mission-critical software;
    - High-performance sensors and
    - Weapon systems.

- **manufacture and repair of AFVs** – There is no absolute requirement to manufacture all of the constituent parts of an AFV in the UK. An onshore capability to repair and overhaul AFVs is however required, both for routine maintenance and in response to operational needs;

- the ability of industry **to respond quickly at times of high operational tempo** is of particular priority. This includes the design and delivery of UORs in a timely manner; the provision of contractor support on deployed operations; to enable a surge repair capability; and the management of a responsive supply chain, which will include the ability to provide secure sourcing of essential raw materials for critical sub-systems. This does not mean that all components must be built on shore, but that the management process to coordinate this complex process must be on shore to guarantee operational availability.

Alvis engineers fitting UORs to Challenger 2 on Op TELIC 1.

Overview of the current defence market

AFV world market overview

B3.18   There is a competitive world market for AFVs. In general, companies that produce sophisticated AFVs in the heavy (30-70 tonnes) and medium (15-30 tonnes) categories rely on their national governments to fund the high development cost of new products and have their national armed forces as lead customer. The cost discourages independent speculative AFV development, for either home or export markets. Once developed, export opportunities are usually limited to nations that have significant investment in their armed forces but no indigenous AFV capability, and are heavily contested. Whilst the UK has had some export success over the last 20 years with the Warrior IFV, Challenger 2, CVR(T) and Saxon, the German Leopard 2 Main Battle Tank and the Swedish CV90 Infantry Fighting Vehicle family have been more successful.

Warrior vehicles of the 1st Battalion, Irish Guards, at a forward mounting base in Former Yugoslav Republic of Macedonia (FYROM).

B3.19   Lightweight AFVs (in the 7 - 15 tonnes bracket) tend to be less sophisticated and both supply and demand are greater and more elastic, with more industry funded product development taking place. However, the entry into this light AFV market of new lower cost manufacturers – from Russia, Ukraine and Turkey for example - means that competition is strong and UK has found itself without a competitive product. We have not developed a significant lightweight AFV since CVR(T) in the early 1970s. Although this was an outstanding export success, it is now generally viewed as obsolescent.

B3.20   The focus which the UK and other nations are placing on medium weight vehicles will offer very significant market prospects over the next 10-20 years, where families of technically innovative, but price-competitive vehicles can be modified to suit national requirements. While the UK currently has no medium weight export product of its own, the FRES solution would clearly be a candidate for such exports. Specialised AFVs such as TERRIER could also be successful exports.

B3.21   The world AFV market consists of several major companies supported by their national governments as outlined. However, the demand for heavy AFVs has fallen below potential supply which has caused recent consolidation. Most notably **BAE Systems** has extended its AFV global presence from the UK, Sweden (BAE Systems Land Systems Hägglunds) and South Africa by purchasing of **United Defense Industries** (UDI), producer of the US Bradley Infantry Fighting Vehicle. The main US provider is **General Dynamics** (GD), which produces the M1 Abrams Main Battle Tank and wheeled medium weight vehicles such as Stryker; GD also has a strong European presence. Both GD and BAE Systems have significant roles in the US Future Combat Systems (FCS) programme. Within Europe, key players include **GIAT**, provider of the French Leclerc Main Battle Tank and VBCI wheeled IFV, and **Krauss-Maffei Wegmann** and **Rheinmetall Landsystems Gmbh** who together co-produce the German Leopard 2 Main Battle Tank. In addition, a number of other suppliers exist in Europe and Asia. There is thus an oversupply of products and capacity in the market, which is driving a move towards consolidation and highlighting the need for increased productivity to drive competitiveness.

UK AFV market overview

B3.22   The main characteristic of the UK AFV Industry in the past 10 years has been its rapid consolidation - from five or more prime companies (GKN Defence, Alvis, Vickers Defence Systems, RO Defence, Marconi Defence Systems etc) to one, **BAE Systems Land Systems**. Drivers for this consolidation include: low profit margins; the significant number of UK national programmes that have not reached product maturity leading to gaps in work load; a lack of competitive export products; a decline in the global export market following the end of the Cold War; and changes in national defence requirements and priorities. However, the industrial model has not changed, resulting in a largely transactional relationship between the Government and industry, supported by traditional post-design service support. Much remains to be done to develop a more modern relationship to manage the AFV capability through life. The consolidation has also resulted in a pressing need for us to develop a strategy with BAE Systems Land Systems in relation to the in-service fleet whilst still retaining access to best of market products at sub-system level.

B3.23   Analysis of the forward programme and the sharp decline in design work in our programmes make it difficult to see how industry can retain the skill base required for the key capabilities identified if we were to continue with our current approach. TITAN and TROJAN are in production, and the TERRIER combat engineer vehicle is in development. These will support existing levels of design capability until the end of 2006, after which there is no assured work. Manufacture of these vehicles finishes in 2010. Export potential exists for TERRIER (and to a lesser extent TITAN and TROJAN) but this is a niche market and, on its own, is unlikely to preserve significant design and manufacture skills in the longer term.

Terrier.

B3.24   The current model of support to the in-service fleet offers opportunities for design and limited manufacture work, but analysis shows that such a reactive and transactional support arrangement does not of itself allow long term industrial planning and restructuring. Nor does the current model sustain the skills needed to carry out technically complex, but infrequent, capability upgrades, including short notice support to UORs.

B3.25   In particular, **ABRO**, a Trading Fund of the MOD provides pivotal support to operations by implementation and support for UORs. ABRO also provide the capability for our armoured fleet base repair and maintenance.

B3.26   Traditionally an AFV consists of a number of major sub-systems, for which the AFV DA defines performance and interface requirements, manufactures them or buys them in, and then integrates them into a working vehicle. In particular, these sub-systems can also enable the integration of the AFV into the wider military network, requiring particular expertise of the overarching electronic architecture. **Thales** and **Lockheed Martin** are currently de-risking aspects of the required electronic architecture for FRES through TDPs, in order to support the role for FRES as a hub of the NEC. Furthermore, AFV combat systems are provided by both **Thales Optronics** and **Selex**.

B3.27   The UK industrial base also has extensive experience in providing the physical architecture for AFVs such as tank track and transmissions provided by the **William Cook group, Caterpillar** and **David Brown Engineering Ltd**.

Sustainment strategy

B3.28   There are two separate but linked areas involved in the sustainment of our AFV capability. The first is improving the through-life management of the in-service capability linked to industrial transformation, and the second is the management of FRES. In addition, we need to understand the linkages between these two distinct but intertwined strands.

B3.29   BAE Systems Land Systems is the DA for 95% of the vehicles in the current fleet. Recognising this reality, we intend to pursue initiatives to change the relationship between us in order that the demands of current operations, routine support and future upgrades are met more cost-effectively. We also seek improved reliability and the reduced deployed footprint necessary to enable the directed logistic approach which is central to the Future Army Structure (FAS). As a company, BAE Systems Land Systems operates in the UK, the US, Sweden and South Africa, but the activities of its constituent businesses are in our view largely autonomous. We wish to encourage BAE Systems Land Systems to become better able to move skills, information and effort within the company to provide improved value for money. We also wish to see continued evidence of a sustained willingness to respond to our defined requirements for improved through-life management of our existing armoured vehicle capability. This improved relationship will require us to take a more coherent through-life view within the Department.

B3.30   This will require real effort from both us and the company. We intend to work with BAE Systems Land Systems to develop a series of clear and incremental steps, which, subject to a satisfactory business case and underpinning commercial arrangements being agreed - would lead us towards a MOD/BAE Systems Land Systems partnering and business transformation agreement. Under this arrangement BAE Systems Land Systems would be incentivised to act as the systems engineer for the current fleet, contracting for capability provision and demonstrating a willingness to exploit the widest possible supply chain in order to benefit from innovation and open competition. This agreement will draw on lessons learned from our collective work under the munitions Framework Partnering Agreement, the Armoured Vehicle and AS90 support initiatives and, more widely, from our work to improve support for helicopters and fast jets. Successful delivery of work on the FV430 and potentially the Warrior and Challenger fleets and the future relationship with ABRO will provide opportunities to build confidence on both sides.

B3.31   We have determined that there is no strategic need to own ABRO, but we also recognise that ABRO provides us with a core capability in the repair and overhaul of the armoured fleet, which must be retained in the UK. We judge that until strategies for the future provision of support for the armoured fleet have matured, any change in ownership of ABRO represents an unacceptable level of risk.

B3.32   For the future, we require industry to deliver an increasingly complex system of systems that will make up the FRES fleet. This includes not only the systems integration of complex sub-systems into the platform itself, which is more than just a question of the physical assembly, but also the integration of the platform into the wider military network. This strategy is likely to involve a strong competitive element. It is questionable whether any single company has the ability or expertise to provide all elements of such a capability, whilst delivering value for money and cost effectiveness. The most likely solution will be a team in which national and international companies co-operate to deliver the FRES platforms, including the required sub-systems, led by a systems integrator with the highest level of systems engineering, skills, resources and capabilities based in the UK. We welcome the international interest in the AFV market and encourage companies to invest in the UK to develop the IP and skills to meet our future AFV requirements. We must have sufficient confidence that we can access the IP, design authority and capability to upgrade or adapt the fleet as required, frequently against demanding timescales.

The technological complexity of AFVs will increase, as evolving threats produce increasingly demanding survivability requirements, and as we seek to realise the benefits of Network Enabled Capability

B3.33   About 30% of our planned AFV research needs to be conducted within Government because of its security classification. We will regularly review the requirement to conduct this entire portion in-house.

B3.34   We currently plan to spend some £140M on research over the next 10 years, in particular in support of the FRES programme. However, some of the non-specific FRES research funding may be directed to specific entities in order to help sustain those key capabilities mentioned earlier. Our current research priorities are focussed on:

- the capability to assess and produce countermeasures to emerging enemy threats;
- survivability – including novel technologies such as lightweight armour systems, active protection systems, electro-optic counter measures, active signature management and integrated survivability systems;
- lethality;
- sensor systems;
- and integration of vehicles into the wider military network.

Callenger 2 tanks of B Squadron, The Queen's Royal Lancers, engage Iraqi Army vehicles on the front line, just outside Basra.

The way ahead

B3.35   We will need to work hard with BAE Systems Land Systems, building on the discussions we have already set in train, and the agreement reached in December 2005, to **give effect to the long term partnering arrangement required to improve the reliability, availability and effectiveness through life of our existing AFV fleets.** Initial activity will focus on implementing measures that build confidence on both sides. We intend to establish a joint partnering team within the early part of 2006 and to establish a business transformation plan underpinned by a robust milestone and performance regime. The plan will detail the improvements in performance to be achieved, the process and behavioural changes required of both BAE Systems Land Systems and the Department, and the capabilities and skills necessary to sustain through life support to AFVs. Under the partnering agreement with BAE Systems Land Systems on the existing fleet we expect to see a significant evolution of BAE Systems Land Systems both to deliver AFV availability and upgrades through life, and to bring advanced land systems' technologies, skills and processes into the UK, drawing on international capabilities. We are particularly keen to see a build up in the UK of expertise in the systems integration of complex land systems. If successful in their evolution, BAE Systems will be well placed for the forthcoming FRES programme.

B3

Armoured Fighting Vehicles

Definition

B4.1   This chapter covers fast jets, air transport, air refuelling, maritime patrol, airborne surveillance, uninhabited aerial vehicles and important aerospace sub-systems.

Joint Combat Aircraft.

Strategic overview

B4.2   Air power remains a fundamental component of war-fighting capability, complementing maritime and ground forces and providing an offensive and defensive capability in its own right. This will be enhanced by the increasing precision of air-delivered weapons and Network Enabled Capability (NEC). Air power continues to offer the ability to transform the battlespace, utilising its inherent attributes of reach and speed to enable strategic operational and tactical agility.   To further enable this agility the RAF is undergoing a transformation and reorganisation in line with the 'Agile Air Force (aAF) concept' to establish agile mission groups capable of rapid reconfiguration to meet dynamic mission requirements. In addition, as an essential part of our future combat air capability, we are examining the balance between manned and uninhabited aerial vehicles. Therefore, whilst there is no current requirement for a new-design manned aircraft beyond our extant plans, future procurements of uninhabited and/or manned platforms are envisaged.

B4.3   The aerospace sector faces a potential watershed as increasing market globalisation, escalating development costs and the absence of any plans for new design manned fast jet aircraft threatens the continued viability of the UK's existing design, development and manufacturing capabilities.

B4.4   This will herald a considerable change both for industry and Government, demanding a significant shift in culture and ways of working. However, there is an enduring need to support and upgrade our existing and planned fleets of manned aircraft, which are likely to have a service life of at least 30 years. Moreover, in order to preserve the ability of the UK to conduct operations without undue dependence on other nations it will be necessary to preserve a number of capabilities onshore.

Equipment programmes

Fast Jets

B4.5   **Typhoon** is a multi-role combat fighter that will replace **Jaguar** (Out of Service Date or OSD 2007) and the **Tornado F3** in providing superior performance and flexibility in both the Air Defence and Strike roles. The two main UK contractors, BAE Systems and Rolls-Royce, have been awarded approximately 37.5% of the total 4 nation work share and are responsible for developing and producing part of the aircraft and engines respectively. The in-Service Date or ISD was achieved in 2003, and the currently assumed OSD is in the 2030s.

B4.6   **Joint Strike Fighter (JSF)** is the planned solution to the UK's **Joint Combat Aircraft (JCA)** requirement which will succeed both the **Harrier** and **Sea Harrier** (which retires from service in early 2006). The JSF will be a stealthy fighter which will be capable of performing multi-role operations from land and sea. The UK is a key partner in the JSF development programme for the US Air Force, US Navy and US Marine Corps and has invested £2bn to date. The expected ISD is in the middle of the 2010s, and the currently assumed OSD is in the 2040s.

B4.7   **Tornado GR4** is a 2-seat ground attack aircraft, capable of delivering a wide variety of ordnance. The GR4a variant provides a low level tactical reconnaissance capability. The currently assumed OSD is in the middle of the 2020s. Looking to the future, we will need to consider replacing the capability currently provided by the Tornado GR4. We are alive to the potential military capability that UCAVs may play in this force mix as a cost effective through life capability.

B4.8   By mid 2007, **Harrier** will be upgraded to **GR9** standard with more powerful engines and electronic systems and able to employ the latest smart weapons in its close air support role. The currently assumed OSD is in the late 2010s.

Hawk Mk1.

Training

B4.9   **UK Military Flying Training System (UKMFTS)** seeks to replace the current Flying Training System in order to deliver the required quantity and quality of aircrew for the three Services, in the most efficient and timely manner. UKMFTS achieved Initial Gate approval in Dec 02, the preferred Procurement Strategy is Public Private Partnerships (PPP), with component parts of the UKMFTS system procured through a mix of Private Finance Initiatives (PFI) and Smart conventional procurement. The **Advanced Jet Trainer** portion of UKMFTS will be met by the procurement of **Hawk 128**; these will replace the current Hawk Mk1 aircraft. A Design and Development contract was let to BAE Systems in 2004. The currently expected ISD is at the end of the decade.

Overview of the current global and UK aerospace market

B4.26   The European and US aerospace market, both civilian and military, is reasonably buoyant at present. Civil aircraft production is expected to increase by more than 40% between 2004 and 2009.

B4.27   The defence market is also entering a new phase of activity with the Dassault Rafale and Typhoon in production and the JSF nearing production. Lockheed Martin is the world's largest defence contractor, with 2004 sales of $35.5 billion and 33% of their sales in the Aeronautic (Aerospace) sector. Some estimate[1] that the worldwide fighter market could be worth some £9Bn p.a. by 2011 with some 30 countries needing to replace aircraft over the next five years.

B4.28   The defence aerospace market, however, is changing. Budgets are focused increasingly on fewer but more capable and flexible multi-role platforms, such as JSF and Typhoon. The trend towards fewer, smaller fleets of ever-more sophisticated, capable and expensive platforms has driven recent changes in the European and US aerospace defence industry. The main consideration for the major suppliers is that there will be a potentially significant reduction in new military aircraft design and development work.

Typhoon.

B4.29   The trend is for collaborative working, either because no single company has the full set of capabilities required to produce world-standard aircraft, or because nations need to collaborate to share costs and each wishes to see some element of the work performed within its territory. This can be clearly seen with both Typhoon (four partner nations: UK, Germany, Italy and Spain) and JSF (nine nations partnering in the development phase: USA, UK, Italy, the Netherlands, Turkey, Canada, Denmark, Norway and Australia). Indeed, the UK has not designed and manufactured a fast jet on its own since the Hawk, first developed in the late 1960s-early 1970s with new variants continuing to be designed indigenously now.

B4.30   Consolidation has become a dominant theme. All tiers of the supply chain have been affected, although at the higher tiers the implications have been more severe. For example, BAE Systems has grown from the merging of 14 aerospace companies within the UK, and EADS combines the capabilities once housed within Aerospatiale, Matra, Dassault (France), CASA (Spain) and DASA (Germany), national companies which themselves were the product of previous mergers. These two companies now dominate the defence fixed wing aircraft market within Europe, primarily via the Airbus Military Company (A400M) and the Typhoon programme. EADS also holds a major shareholding in Dassault Aviation, the manufacturer of Rafale. Similarly, BAE Systems holds a 20.5% stake in SAAB AB, the manufacturer of the Swedish Gripen aircraft.

---

[1] *Source: The Teal Group*

B4.31   There has been similar consolidation in the defence electronics sector. Following a series of agreements struck with BAE Systems, Finmeccanica has become Europe's second biggest operator in the defence and security electronics sector, and the world's sixth biggest through its three SELEX companies[2], and Smiths Aerospace has developed into one of the leading transatlantic aerospace equipment and systems companies, with more than 10,000 staff and nearly $2bn revenues split between Europe and North America.

B4.32   The role of prime contractor is also changing to one focused more on development of systems architectures facilitating insertion of other suppliers' sub-systems, rather than vertically integrated models. We will need to rely increasingly on specialist contractors for individual sub-systems and components for the insertion of new technologies and capabilities into our aircraft fleets. The major prime contractors are generally, by nature, defence companies and must remain focused on defence markets. They may have adjacent businesses in the civil sector, for example defence and civil aerospace interests, but these are largely driven by different business models. Further down the supply chain there is more scope for leveraging common technology and capability between civil and defence sides of the business and some businesses have highly diversified and often international portfolios. This implies that prime contractors may be more likely to face sustainment issues.

B4.33   Rolls-Royce is one of the world's largest military aero engine manufacturers following a number of acquisitions including the US Allison Engine Company and BMW's aero-engine division. The US companies General Electric and Pratt & Whitney and the French company SNECMA are now the other major players in the industry.

A C-130 Hercules lands at the austere airstrip of Archers Post in Kenya.

B4.34   The US market, driven by larger defence budgets and greater scale of production, will offer increasingly important opportunities for those aerospace companies able to gain access to US programmes. Both the major European primes, BAE Systems and EADS have opportunities in this arena, the former due to its presence in the JSF programme and its increasing US focus (bolstered by US acquisitions) and the latter potentially via its teaming with US primes on current US programmes. Rolls-Royce is also playing a key role in taking forward the JSF programme. Indeed, their contribution is central to the STOVL variant.

B4.35   The future aerospace market might show a significant shift to the widespread use of uninhabited platforms, a field in which US and Israeli companies have notable experience. European companies, including BAE Systems and EADS, are also active in this field, with all seeking to develop competitive advantage in what appears to be a growth area.

---

[2] *SELEX Sistemi Integrati, SELEX Communications and SELEX Sensors and Airborne Systems*

Sustainment strategy

B4.36   The issues facing the Aerospace Sector are a **matter of mutual concern to Government and Industry alike**. The Defence Industrial Strategy gives timely context for this.

B4.37   At one level, we need to be clear that both the nature and volume of customer demand are changing, with implications for the industrial base. **The coming decline in new programme work will have an impact on the UK industrial footprint**, in particular around BAE Air Systems' four main production sites (Warton and Samlesbury in Lancashire, Brough in the East Riding of Yorkshire and Woodford in Cheshire). The challenge is to manage this transition and sustain in the UK – in the absence of major new programmes – the industrial skills, capabilities and technologies that are required to sustain our ability to operate, support, maintain and upgrade our aircraft over the next 30 years. **MOD has been working closely with BAE Systems, as the UK's only supplier of fast jets, for some time to understand these mutual challenges**. We are committed to continuing this dialogue with a view to finding a solution that meets the defence needs, now and in the future, recognising that this will need to make commercial sense for the company.

B4.38   This is likely to demand a long-term and strategic approach. To that end, MOD and BAE Systems intend to work together to explore how a long term partnering arrangement for the through-life availability of a significant proportion of the fixed-wing fleet might be delivered. Such an approach, if delivered, would allow rationalisation (to take out surplus capacity), improved efficiency and better ways of working, capacity management (to smooth workloads), exit and step-in provisions, and open-book accounting; there has to be alignment of objectives. Specifically, incentivised platform availability and in some cases platform capability contracts with industry will be crucial to the future delivery of operational output at the frontline. It will demand changes on both sides to conduct business differently, and is probably best achieved through a phased approach so that the risks can be progressively tackled - and success, for both parties, assured. This needs to extend through the entire aerospace supply chain, as all have a part to play in delivering this capability through-life.

B4.39   Our need to retain a minimum level of onshore capability does not necessarily mean that we will need to support all aspects of our aircraft in the UK. This is because generic capabilities and skills developed to support one aircraft type may be brought to bear on another, depending upon similarity of type, role, technology and complexity. In considering whether to retain an onshore support capability for any particular aircraft fleet, we will first determine where the best value for money can be obtained. Where this is offshore, we will assess any implications of pursuing that route on our ability to preserve a minimum capability across the breadth of our business. Thus, at the strategic level, we will manage what amounts to a portfolio approach.

B4.40   For any particular aircraft type there may also be a middle ground where, to secure value for money for example, we may rely on off-shore suppliers for major upgrade but retain onshore maintenance, support and Urgent Operational Requirements (UORs) capabilities.

B4.41   For **Typhoon**, we will design a cost effective and affordable national support solution, incorporating best practices from lean and end-to-end principles. We intend to work with our quadrinational partners to effect this. However, major changes must be made in the management and operation of the supply chain to incentivise continued improvements in the support arrangements for this aircraft, ensuring that we retain onshore our ability to satisfy our sovereign requirements over its lifetime. Clearly, BAE Systems, and, for the engines and mission systems respectively, Rolls-Royce , Smiths Aerospace and Selex Sensors and Airborne Systems will have a significant role to play in this.

B4.42   For **JSF**, the through life support of the UK aircraft will be provided from the Lockheed Martin Global Support System which is being established on a co-operative basis amongst the 9 JSF partner nations. This will provide support, through performance based contracts, to the JSF fleet. As part of this performance based arrangement, the UK also intends to establish sovereign support capabilities which would provide, inter alia, in country facilities to maintain, repair and upgrade the UK fleet and an Integrated Pilot and Maintainer Training Centre. Our aim, endorsed by the US Department of Defence (DOD) and agreed with Lockheed Martin, is that BAE Systems as a key JSF Industry partner to Lockheed Martin will provide these support services in the UK under a Team JSF badge with contracts flowing from the MOD to DOD to Team JSF. There is no fundamental defence requirement for a JSF Final Assembly and Check Out (FACO) facility, although an on-going joint study with DTI/BAE Systems, due to conclude in early 2006, is seeking to assess whether a UK FACO is necessary to preserve essential engineering skills within BAE Systems and would be a cost effective solution or whether alternatives would provide a better outcome in terms of sustaining core skills.

Propulsion

Power systems are central in the air, as well as at sea. In this field, Rolls-Royce are a champion within the UK defence industry, and a world leader in aero-engines, marine and land systems. 100% of our major warships and 80% of our aircraft are powered by Rolls-Royce engines.

We will wish to retain in the UK the ability to support and upgrade our platforms through-life, and sub-systems, including engines- often extremely complex in their own right- are often the key route to improving reliability and other aspects of capability. Rolls-Royce is already playing a significant role as a strategic partner to the MOD in this area- as demonstrated by the innovative through-life engine support arrangements already in place for Harrier, VC-10 and Nimrod.

We will also continue to invest in propulsion technologies where these show potential application to our future needs. New work includes the Affordable Combat Engine Technology TDP, in which we will invest about £12.4M over 4 years.

B4.43   We will also engage closely with our major suppliers to review the applicability of **civil standards** to military aircraft. This will allow us to gain maximum leverage from the UK civil aerospace market and to focus our resources more specifically on sustaining capabilities that are unique to the military environment.

Aerospace systems and enabling skills and technologies

B4.44   Further work to identify strategically important underpinning skills and technologies will be undertaken, which we hope will be completed by Autumn 2006, and we will work closely with industry to understand how these might be retained onshore. Our general strategy will be to target more accurately our existing research budget, possibly through research partnerships. This will help UK industry to identify and exploit new techniques and technologies and to remain at the technological forefront of new developments. We do not envisage any significant investment in production facilities; rather we expect to benefit from downstream economy of scale savings as a result of military export sales and commercial exploitation.

Uninhabited Air Vehicles (UAVs) and Uninhabited Combat Air Vehicles (UCAVs)

B4.45   We and industry share a close alignment of interest in UAV and UCAV technology. Although at present we have no funded UCAV programme, targeted investment in UCAV technology demonstrator programmes would help to sustain the very aerospace engineering and design capabilities that we need to provide assurance of our ability to operate and support our future fixed wing aircraft. Such investment would also ensure that we can make better informed decisions on the future mix of manned and uninhabited aircraft which will need to be taken in the 2010-2015 timeframe. Additionally, the benefit for UK industry is the opportunity to develop a competitive edge in a potentially lucrative military and civil market.

B4.46   In the context of the wider discussions with the industry around consolidation and transformation, we are considering ways in which we can take such an aspiration forward. BAE Systems is leading a UK industry team working on UAV technologies, following some recent very successful company and MOD-funded technology demonstration programmes. This work has pioneered a range of agile project management techniques; an absolute focus on key objectives, a fast decision making process, and rapid prototyping and engineering. This approach, which we are keen to use more widely, has significantly cut the time in which new ideas and technologies can be realised and demonstrated. For example, BAE Systems' own Raven UAV went from concept to first flight in ten months. Building on the success of these programmes, we intend to move forward subject to a value for money business case being demonstrated and appropriate commercial arrangements being in place with a more substantial TDP (Technology Demonstrator Programmes) designed to give us and industry a better understanding of key technologies of relevance to UAVs and UCAVs more broadly. This would be a joint effort with MOD and industry contributing to the costs. We hope that appropriate arrangements will be in place to allow this activity to proceed in 2006.

The challenge ahead

B4.47   The aerospace sector is facing a tough challenge. The UK has not had the capability to design and manufacture on its own the most advanced combat jets for a long time, and yet the principal onshore supplier of fixed wing military aircraft has four sites dedicated to air systems. The current size of the air sector is not sustainable, and rationalisation and reduction in terms of both infrastructure and employment is inevitable.

B4.48   The MOD recognises the difficulties that reduction in planned new programmes is likely to entail, and we will work with BAE Systems and the other companies in the defence aerospace sector so that it can reach the appropriate size and shape for the demand. With this process, we can help the sector to remain a healthy, competitive and profitable one that can survive into the long term to meet our changing future requirements. Without it, the inevitable reduction in the global defence aerospace industry will happen anyway, but in a less structured and ordered way and not necessarily with the UK at the forefront - to the detriment of the companies and their employees.   But if we, MOD and industry, face this challenge up-front, together, we can ensure that the UK is well positioned for the future.

B4.49   Our plans to retain onshore the industrial capabilities required to ensure effective through-life support to the existing and planned fast jet fleet - and to invest in developing UCAV technology - will also provide us with the core industrial skills required to contribute to any future international manned fast jet programme, should the requirement for one emerge. This recognises both the uncertainty of our very long term requirements - with the possibility that we shall want to replace elements of the Typhoon and Joint Strike Fighter fleets with manned aircraft - and that we should avoid continuing to fund industrial capabilities for which we have no identified requirement.

The way ahead

B4.50   We need to develop the dialogue in which we have been engaged by **commencing negotiations with BAE Systems in earnest on the terms of the business rationalisation and transformation agreement required to facilitate the effective sustainment of the industrial skills, capability and technologies – wherever they may be in the supply chain – that will be so important to our ability to operate, support and upgrade our fast jet combat aircraft through life**. We aim on working with the company during 2006 to agree the way ahead – which will be challenging given the scope of the scale of the transformation that is required – and to implement it from 2007.

Definition

B5.1   This chapter addresses helicopters and those systems that are unique to them; other avionic systems are addressed in the Aerospace chapter.

Strategic overview

B5.2   Helicopters play a major role in the UK's military operations. The Battlefield Helicopter (BH) fleet has recently operated in a wide variety of theatres, including urban and rural areas in Northern Ireland, the Iraqi desert, the mountains of Afghanistan and the jungles of Sierra Leone.

Chinook Mk2 transporting a Light Gun during Op SILKMAN Sierra Leone.

B5.3   Helicopters are inherently responsive, adaptable and flexible, and contribute to a variety of military tasks. They can operate in a very wide range of combat and environmental conditions. Therefore to be an effective component of a balanced expeditionary force they need to have high reliability and availability and require the minimum possible deployed logistic support equipment and infrastructure.

B5.4   **The Future Rotorcraft Capability (FRC)** programme was created in July 2004, to identify a future strategy that maximised the capability that could be delivered from available funding. The FRC programme was directed to explore opportunities to use each helicopter type to deliver more than one capability, reduce the number of types of helicopters in-service and promote off-the-shelf (OTS) solutions, limiting unique UK requirements to the essential in order to drive down costs of ownership. To understand better the nature of future helicopters requirements, a FRC taxonomy was agreed and used to define the overall capability, recognising that individual platforms contribute routinely across the capability domains. For combat helicopters, three capability domains of **Attack, Find** and **Lift** were identified and, within each, separate environmental requirements for **Land** and **Maritime**:

Capability Domain	Description
Attack	A helicopter capable of autonomous and co-operative attack using appropriate weapons against surface (land and maritime) and sub-surface targets
Find	A helicopter capable of autonomous action, which provides tactical commanders with reconnaissance, surveillance and target acquisition
Lift	A helicopter capable of a vertical-lift capability to support military operations, which must enable the rapid deployment, in-theatre movement, re-supply and extraction of joint forces and their equipment

B5.5   The requirement to move ship-based troops and equipment to the onshore Battlespace is known as **Littoral Manoeuvre** and is captured in the Land elements of the taxonomy.

B5.6   A number of early priorities have been agreed during an initial phase of analysis to drive the delivery of improved capability in the early years. These included: sustaining the capability currently provided by Land and Maritime Lynx and Gazelle aircraft; securing a value for money approach to modernising the maritime Merlin Mk1; progressing with a helicopter-based replacement search and rescue (SAR) capability; and undertaking restorative measures in Lift. Work necessary to define how much, and when, to invest further in the Land Lift capability, including the balance between Medium and Large Lift aircraft, will be undertaken in the **Land Advanced Concept Phase (LACP)**, reporting in late 2006.

B5.7   Search and Rescue (SAR) activities in the UK, Falkland Islands and Cyprus are treated as a separate capability area, as are basic helicopter training needs (provided by the Defence Helicopter Flying School), support to training activities and general liaison roles.

Equipment programmes

B5.8   The **Merlin Mk1** Capability Sustainment Programme (CSP) aims to ensure continuity of capability and introduce an open-systems architecture to the Royal Navy's (RN) airborne Anti-Submarine Warfare (ASW) capability. It will also enable the cost-effective management of obsolescence on an aircraft which has components and design features that are becoming difficult to support. This programme is scheduled to achieve its In-Service-Date (ISD) in the middle of the next decade. Merlin Mk1 is likely to remain in-service into the 2030s.

B5.9   Replacements are needed for capabilities provided by the Lynx and Gazelle aircraft in service with both the Army and the RN. The preferred solution, currently undergoing detailed analysis to ensure, amongst other aspects, its value for money, for these requirements is the **Future Lynx** performing as the Army's **Battlefield Reconnaissance Helicopter (BRH)** capability and the RN **Surface Combatant Maritime Rotorcraft (SCMR)**, although some Gazelle will be retained in the non-combat training, support and liaison roles. Both BRH and SCMR are scheduled to be delivered in the middle of the next decade.

Interior of Merlin helicopter of 824 Naval Air Squadron.

B5.10   A package of work known as the **Chinook Mk2/2A coherence programme** has been launched to establish a single configuration baseline for the Chinook fleet and is expected to be completed early in the next decade. This work will enable a reduction in the overall cost of supporting a fleet with disparate equipment standards due to the fitting of a large number of partially integrated equipments in support of operations. It is also an essential precursor for the integration of the Bowman communications system and will enable a future **Chinook** capability sustainment programme, should it be decided to extend the life of the platform. This work is expected to complete around the turn of the decade. Assessments are also well underway on the work necessary to field the 8 **Chinook Mk3s** procured from Boeing in the late 1990s which are not in service as we have been unable to certify their airworthiness. A decision will be taken next year on whether to proceed.

B5.11   Currently, the majority of **Search and Rescue (SAR)** helicopters in the UK are provided by the RAF and RN, with the remainder provided by civilian helicopters under contract to the Maritime & Coastguard Agency. It is planned to begin to replace this capability with a single contract that retains a proportion of military aircrews to enable operational readiness of Combat SAR crews in the middle of the next decade.

B5.12   The LACP will make recommendations on our future **Lift** capability and the appropriate balance of investment on how this should be taken forward. Key to this activity will be determining the balance between investing in new equipment or sustaining and enhancing present lift helicopters. The Puma, Sea King Mk4, Merlin Mk3 and Chinook Mk2/2A will all need investment if their contribution to the Lift capability is to be extended beyond the middle of the next decade, since all will reach their planned OSDs or require obsolescence issues to be addressed within this timeframe.

Apache.

B5.13   In addition, the existing fleet of 67 **Apache Attack Helicopters (AH)** will continue to meet the Attack role in the Land environment until its currently assumed OSD in around 25 years. Work is ongoing to determine how the AH can be sustained through life, including investigations to extend its potential OSD by another 10 years. This will be co-ordinated with the US Army programme to reduce costs and produce appropriate commonality.

Indicative planning assumptions

B5.14   Figure B5(i) shows the current profile of funding assumed for all helicopters within the Equipment Programme (EP). We currently expect to spend an average of £600 million a year on support costs over the next ten years. Our goal is to reduce the cost of ownership by: exploiting new airframes; innovative support arrangements; improved commonality of equipment training and support solutions; and streamlining acceptance and release-to-service arrangements.

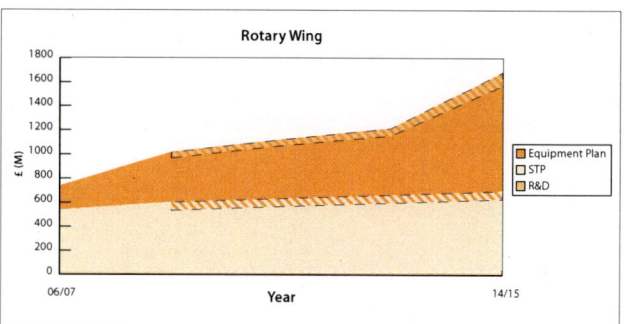

Figure B5(i) Illustrative spend profile.
The above graph shows indicative spending in this sector over the next ten years. The figures from 08/09 are illustrative and include a range in order to emphasise the potential for shifts in investment priorities after the end of the current Spending Review period. This is prudent planning which does not distort the overall illustrative picture of general trends.

B5.15   The MOD has various helicopter Private Finance Initiatives (PFI), which sit outside the FRC and AH programmes, which the Joint Helicopter Command (JHC), the RAF and the RN use to provide aircraft for the Military Flying Training School (MFTS), training support, surveillance, VIP flights and other non-operational liaison tasks, mainly in the UK, but also in Belize, Brunei and Cyprus. These contracts amount to over sixty aircraft, which are Civilian Owned Military Registered (COMR) and equal spend of about £42M per annum to industry. It is also likely that a COMR-type solution will be sought in the future to replace the training support and non-combat liaison tasks currently undertaken by Gazelle. In addition, there may be scope to combine some or all of these contracts in due course.

What is required for retention within the UK industrial base?

B5.16   **Support of current aircraft** – The retention onshore of those skills critical to the through-life support of our current UK designed aircraft is essential, in particular to ensure the airworthiness of the platform. This includes modification and programmed upgrades, which typically includes the provision of new sensors and defensive aids, structural repair and the urgent insertion of new capability in direct support of ongoing operations. These skills are primarily resident in AgustaWestland, the original manufacturer of much of the in-service fleet, and will need to be sustained to ensure our current aircraft can be supported.

B5.17   **Systems engineering** – In order to address the demands of the future network-enabled battlespace a broad spectrum of systems engineering skills will be required, not least to help ensure the support of our existing aircraft can be undertaken. Without these skills we may not be able to upgrade and insert new technologies into the existing fleet. These skills range from the integration of platform, powerplant, navigation and communications systems through to the more complex integration of mission system, sensors and processors. We would also wish a modelling and simulation capacity to be retained within the UK. We expect that these skills will be sustained through ongoing helicopter acquisition and upgrade programmes and other major Defence activities outside the helicopters environment. For example, Merlin CSP will help maintain and enhance those required skills at Lockheed Martin in Havant (the Design Authority for this system) and Thales UK (as a significant equipment provider). We would also wish to exploit the wider capacity available to multinational companies, not least by managing workloads more efficiently across the breadth of those companies.

Key technologies to enhance future platform development

B5.18   As helicopters cover a wide span of technologies and enabling areas of expertise, the totality would be unaffordable to sustain through our demands alone. However, there are particularly important technologies for helicopters within niche areas, which provide the Armed Forces with unique capabilities and offer leverage in our co-operation with allies and the broader design and manufacturing base.

B5.19   **Rotor blades** - Rotor blade technology is a key area of UK expertise that has, for example, allowed the 15 tonne Merlin Mk1 aircraft to operate within the same maritime rotor diameter constraint as the 10 tonne Sea King variant which it replaced. The UK has been instrumental in investigating performance and enhancing blades. Looking ahead, this technology offers better range and payload capabilities and could lead to lower whole life costs and vibration. Investment in the British Experimental Rotor Programme (BERP) technology will continue until the technology has been matured and successfully demonstrated in flight. This will sustain the skill base, although decisions about integration of BERP IV blades onto our helicopter fleet will be made on their merits.

B5.20   **Mission systems** – The threat spectrum in which helicopters operate places a premium on situational awareness, augmented by on-board decision-making aids and enhanced Human Machine Interface (HMI) integration, such as speaker-independent voice control. The design and manufacture of compact, world-leading sonar and radar systems is an example of where the UK plays a leading role in helicopter-specific mission systems. Continued investment in the COvert Night Day Operations Rotorcraft (CONDOR) technology development programme, which could offer exploitation opportunities around the middle of the next decade, will help to sustain this skill base.

B5.21   **Survivability** – Helicopter Survivability requires a systems approach across a broad spread of capabilities including those Electro-Optical (EO) sensors, Radar, Electronic Support Measures (ESM) and Defensive Aids Systems (DAS) essential to underpin the UK's land and maritime capabilities. It is a sensitive and critical capability that provides choice to the commander in the pursuit of his mission. We intend to protect the security of supply of these underpinning technologies and to minimise the opportunities for our adversaries to counter those systems through knowledge of their operation. We will retain the ability to manufacture prototypes, test and evaluate potential solutions and use crew-in-the-loop simulation to develop tactics to support ongoing UK operational commitments.

B5.22   **Vibration management** - The UK's ability to design, prototype and validate active vibration management systems offers significant military and cost of ownership benefits. The ability lies in a physics-level understanding of how vibrations are transmitted through the helicopter and addressing this through active systems; this is in contrast to other nations, which rely

heavily on trial and error methods. The technology enables members of the Armed Forces to remain operationally effective in an otherwise disruptive environment caused by the vibrations from the platform. It has been fitted onto Merlin Mk1. We will sustain this expertise through continued research, linked to improved exploitation with industrial partners.

HMS ARK ROYAL in the Northern Arabian Gulf, is resupplied by a Royal Navy Sea King of 820 Naval Air Squadron.

B5.23   **Electronic architecture** – There are a variety of electronic systems which are vital to both the safe flight of helicopters and also their ability to fight on the battlefield. As a minimum we would wish to retain the knowledge of the controlling and source software within such capabilities as:

- Infra Red (IR) systems
- Electro-Optic (EO) sensors
- Mission management software
- Electronic Warfare (EW) systems

Overview of the current global defence market

B5.24   **Market background** – The five major manufacturers in the global helicopter market are Boeing, Sikorsky, Eurocopter, AgustaWestland and Bell. Each harnesses the majority of its revenue from military sales, which are predicted to grow by 60% over the next decade, as a result of the US helicopter programme. The civil market is likely to remain relatively static in comparison. This represents a significant opportunity, following recent success in the US Presidential helicopter replacement programme, for AgustaWestland as both Future Lynx and Merlin have excellent export potential, particularly in the maritime segment of the future helicopter market.

B5.25   Reductions in defence spending in all European Union countries has driven a marked fall-back in the amount of funds available for the research and development of new capabilities. As a consequence, further integration is likely to be required in the European market, and access to the US market will be critical to the viability of all helicopter manufacturers. We believe the UK has an important advantage due to its close links with the US.

Overview of the current UK defence market

B5.26   The UK helicopter market is dominated by **AgustaWestland**, who currently provide the in-service and through-life support to the majority of the UK's helicopter fleet. AgustaWestland form part of the Finmeccanica group of companies.

B5.27   **Lockheed Martin UK**, part of the wider Lockheed Martin group, is the prime contractor for the RN Merlin Mk1 fleet of aircraft.

B5.28   The majority of our helicopter repair and overhaul is undertaken by the **Defence Aviation Repair Agency** (DARA), a trading fund of the MOD. The Department has recently announced that it will consider taking to the market DARA's rotary wing and associated components in order to test whether their sale might deliver improved effectiveness and value for money for the Armed Forces.

B5.29   UK helicopter sub-contractors include **Thales UK** who are a major supplier of avionics and other sub-systems for helicopters such as mission systems and radios and are also the prime contractor for the conversion of the Sea King ASaC Mk7 aircraft. **Rolls-Royce Defence Aerospace** support the GEM, GNOME and RTM322 (in conjunction with TurboMecca) helicopter engines. **Westland Transmissions Ltd** design and develop helicopter transmission systems and **Smiths Industries** provide a variety of electronic and flight control systems. Both **General Dynamics UK** and **Selex Sensors and Airborne Systems** provide avionics and information management systems for helicopters. AgustaWestland and **QinetiQ** also provide particular capability in devising the Vibration Management technology outlined above.

Merlin operating in Iraq.

Trends

B5.30   The helicopter industry and its sources of revenue are changing, increasingly moving away from the provision of equipment to contracting for services and involving industry in the direct support of operations.

B5.31   Contracting for Availability, Integrated Operational Support (IOS) and Through-Life Customer Support solutions will require a number of industrial competencies to respond to helicopter capability requirements. These new arrangements do not significantly change the technical demands to support the platforms, but they do generate a requirement for industry to transform to a new business model where industry is incentivised to improve helicopter availability and reliability through-life.

Sustainment strategy

General

B5.32   Within the context of our agreed FRC strategy, our sustainment strategy seeks to develop a closer, more transparent relationship with the helicopter industrial sector in order to deliver a more coherent and cost-effective through-life capability. We intend to sustain a strong systems engineering capability, although it is not an absolute sovereign requirement to maintain a separate national helicopter design and production capability within the UK for new aircraft. However, we will actively encourage appropriate partnering arrangements with industry where it makes sense in terms of better value for money, maintaining appropriate sovereignty over our helicopter capability and in enabling business transformation. We would expect such agreements to draw on the breadth

and depth of the industry's capacity, both in the UK and from abroad.

B5.33   In managing better our current fleet and the transition to the future we intend to develop further our IOS arrangements towards a common approach with industry in meeting our broad capability needs over time. As we recently announced, we would hope the envisaged sale of DARA Fleetlands will provide the opportunity to create a more robust and profitable support base in the UK. Finally, we will encourage the development and exploitation of key niche technologies, acting where appropriate in close co-operation with our allies and industry.

Sustainment of the existing fleet

B5.34   Sustainment of the existing fleet entails the provision of necessary support arrangements and an ability to respond to airworthiness issues, crash investigations (including the delivery of any remedial action) and the effective and timely embodiment of capability enhancements, sometimes at very short notice. This requires a wide spectrum of engineering skills (including structural design, aerodynamics, dynamic systems, avionics integration, powerplant integration, test and evaluation and software design), knowledge of UK military demands and safety standards, plus competence in the integration of sub-systems into aircraft. These skills are sometimes collectively referred to as a Design Authority (DA) capability.

AgustaWestland

B5.35   For the majority of our existing aircraft, AgustaWestland have a DA role: for Lynx, Sea King and the battlefield variant of Merlin they are the platform and systems DA; they are the platform DA for the RN Merlin; for Apache, they have a role to co-ordinate inputs of platform and systems; and for Puma and Gazelle, they are the DA for UK specific systems. The sustainment of these skills requires a sufficiency in each discipline to ensure a depth of knowledge, experience and effective succession management. **The professionalism and effectiveness of this skill-base is best sustained by ensuring that individuals are faced with 'high tech' engineering challenges, typically only found in helicopter design, development or demonstration activities**.

B5.36   The AgustaWestland skill-base is at present exercised through export opportunities and sub-contract work on Merlin Mk1 and the US Presidential Helicopter, but this is insufficient to sustain the breadth and depth of engineering skills required nor support the transformation of the company. Therefore, to meet the BRH and SCMR requirement **our preferred solution is to invest in the Future Lynx product**, currently undergoing detailed capability and value for money assessment, as it appears to provide the required military capability and also sustains the necessary DA capacity at the company in the short to medium-term.

B5.37   We also intend to **promote a more open, predictable but demanding partnered relationship with AgustaWestland, to realise business transformation within the company. We wish this transformation to provide better value for money and reduce their reliance on our investment to sustain this design engineering skill-base**. This work will also assess how partnering will help drive coherence in, and value for money from, the future support arrangements for our current helicopters and other contracts placed with AgustaWestland. **We intend this process to be both challenging and as open as commercial confidences will allow.**

Integrated Operational Support

B5.38   **We have already started work with a number of helicopter suppliers to implement revised and novel arrangements to support our current platforms through long-term, partnered** contracts that require industry to provide serviceable aircraft at the front-line. An Integrated Operational Support (IOS) approach for Sea King is in place and similar approaches are being considered for other aircraft. This is already delivering improved performance in aircraft sustainability and availability, together with associated management information to facilitate performance management. At present, tailored arrangements have been pursued on an aircraft-by-aircraft basis, but we intend to focus efforts on ensuring these initiatives converge and that lessons are learnt, shared and implemented.

Humanitarian aid is off loaded from a RAF Puma in Mozambique.

Wider industrial engagement for the future

B5.39   We will explore how partnering contracts and concepts can be utilised in areas of the helicopter sector. **But it is important to be clear that we will continue to look to the vibrant and competitive global market place to satisfy our future helicopter requirement with AgustaWestland's role neither predefined nor guaranteed, but dependent on their performance and the value for money of their propositions.** The support arrangements for our future helicopter fleets, irrespective of their source of supply, will also be determined on value for money arguments (recognising that it will be important to be clear that they are coherent with existing support infrastructure and capabilities).

B5.40   We also plan to work with Boeing to improve the through-life support arrangements to our Chinook fleet. This supports our aspiration for effective capability management for current aircraft. Boeing, as the DA for the Chinook aircraft, is best placed to undertake the necessary logistical transformation needed in the supply chain that ensures Chinook can be upgraded when we wish in order to develop military capability. Much of the work will be spilt between DARA for depth maintenance and the front-line base to ensure a viable onshore presence.

Exploiting research

B5.41   We will continue to invest in research in support of the development of key helicopter related technologies. This investment averages approximately £13M per year and it is anticipated that it will remain at this level for the foreseeable future. Our priorities include: enhanced helicopter survivability; operations in Day/Night and all weather and adverse environmental conditions; improved mission decision support systems; research into improving mission performance; and ways of reducing whole life costs.

The way ahead

B5.42   We need to **drive forward with AgustaWestland the implementation of the business transformation partnering arrangement** that we committed to through the Heads of Agreement signed in April 2005. A partnering team (jointly resourced by MOD and AgustaWestland) has, for the last six months, been exploring partnering opportunities across those areas of the business indicated in the Heads of Agreement. **We hope that by the Spring of 2006, subject to value for money having been demonstrated, we will have reached agreement on a Strategic Partnering Arrangement (SPA)** which will be focused on activities to sustain the design engineering skills and knowledge of UK military demands and safety standards within the company necessary for them to provide effective through-life support to those elements of the in-service helicopter fleet for which they are the DA. The SPA is intended to commit both parties to specific targets on, inter alia, cost and schedule adherence (and where appropriate improvement) and improvements in operational availability; it will be underpinned by a Business Transformation Plan that sets out the process and behavioural changes required both by MOD and AgustaWestland.

Definition

B6.1   General munitions are those "simple" munitions that do not tend to require interventionist maintenance procedures. The technology is based primarily on energetics (explosives chemistry and mechanical engineering) and where intervention is necessary it is simple and requires generic engineering capability. They do not require to be managed with their parent systems

Strategic overview

B6.2   General munitions, be they Land, Sea or Air delivered, provide an essential part of the defence arsenal. Notwithstanding the drivers for increased accuracy and lower stockpiles, there will be an enduring requirement for general munitions for the foreseeable future to complement an increasing arsenal of complex weapons.

B6.3   Recent operations have clearly demonstrated that despite the increases in technology, modern warfare, particularly on the ground, requires highly trained and motivated service personnel to engage in combat at a very personal level. It is in such engagements that quality general munitions, including Small Arms Ammunition (SAA), are essential to provide the volumes of fire and the 24 hour, all weather capability required to suppress, neutralise and demoralise enemy forces. Security of supply for general munitions will therefore remain an important consideration.

Equipment programmes

B6.4   The major influence on future general munitions programmes is our policy on **Insensitive Munitions (IM)**. The energetic materials that give munitions their propulsive and destructive power also make them susceptible to accident and combat stimuli such as heat, shock and impact. A number of design techniques exist which can reduce munition vulnerability, and minimise the potential collateral damage from an event. Munitions incorporating these design features are termed IM. Under UK legislation we have a duty to ensure, as far as reasonably practicable, that our munitions are designed and constructed to be safe and without risk. As such, all future munitions must be IM compliant. Where existing non-IM compliant munitions cannot be converted, all new buys of these munitions will require an IM Waiver. The aspiration is that by 2010 all new UK munitions will be IM, although some legacy non-IM systems may remain in service beyond 2010.

During Operation TELIC, of the 4,400 containers initially shipped to the Gulf, over 25% contained munitions – some 20,000 tonnes

B6.5   The **Tubed Artillery Conventional Ammunition System (TACAS)** programme is looking at the future requirements of the **155mm family of munitions.** This may include a more efficient and effective charge system, improved projectiles, and new packaging, asset tracking and smart tagging. The **Fuze and Fuze Setter (F&FS)** programme seeks to deliver a F&FS for the 105mm/155mm suite of munition and will be available to all platforms.

B6.6   The **105mm Improved Ammunition (IA)** project is procuring an improved High Explosive (HE) round (L50) for use with the Light Gun. The L50 is more cost effective and lethal than the in-service L31 round, and has a much improved IM signature. The new shell will use existing fuzes until the introduction of the IM TACAS F&FS towards the end of the decade. Currently the assumed ISD is 2007.

B6.7   **4.5" Improved Ammunition (IA)** is in service but not yet fitted 'fleet-wide' across the Royal Navy (RN). IA increases the range of 4.5" ammunition; a separate programme is investigating the potential to move to a fully compliant IM round. A study is exploring options for improving the capability of the RN's medium calibre gun.

B6.8   The current OSD for **CRV-7(FW)** is around the end of the decade to coincide with the projected OSD for Harrier GR7.

AS90 firing High Explosive ammunition.

B6.9   The **L27 CHARM3** is the current in-service 120mm **Armour Piercing Fin Stabilising Discarding Sabot (APFSDS)** Depleted Uranium (DU) tank round. The OSD is not yet known but is dependent on the shelf lives of the L16 and L17 charges, which are being evaluated. A possible Low Vulnerability (LOVA) replacement charge is being studied. For APFSDS training, the **CHARM3 Training Round (C3TR)** – comprising the inert L29A1 (shot) and the L18A1 (charge) – is used. The replacement of the **L23A1 APFSDS** tungsten round, which has an OSD of 2008, by a new tungsten round is being investigated. For the longer term, research is ongoing to assess the option of replacing the current rifled barrel with smoothbore. If this option were taken it is possible that, although smoothbore ammunition is currently available off the shelf, UK would seek to develop, potentially with overseas partners, new ammunition in order to match emerging threats that defeat current smoothbore ammunition.

B6.10   The **L31A7 High Explosive Squash Head (HESH)** OSD is currently assumed to be around the end of the decade. **L32A6 Squash Head Practice Round (SH-PRAC)**, an inert HESH training round has no IM issues and hence no planned OSD. The 30mm Rarden cannon fires the **Armour Piercing Discarding Sabot (APDS)**, which has a currently assumed OSD around the end of the decade. The **30mm High Explosive (HE) round** is to remain in-service until 2012, it is not due to be replaced.

B6.11   The ammunition used by:

- the **sniper rifle** and **long range rifle** have expected ISDs later in the decade and are likely to run for 15 years in alignment with the weapons;
- the **0.5 Machine Gun** and **Barrett Anti Material Rifle** OSD is 2015 in alignment with the weapons;
- **Automatic Grenade Launcher** has an ISD is 2007 and has a shelf life of 10 years;
- the **51mm Light Mortar** has an OSD of 2007 for smoke and illumination natures and 2010 for high explosive. 51mm capability may be replaced by **40mm grenades**;
- the **81mm mortar** is planned to be in service for at least the next 15 years.

SA80 with the Underslung Grenade Launcher.

B6.12   The **Next Generation Light Anti-Armour Weapon (NLAW)** is a short-range shoulder launched unguided system with broad applications across all arms and services and will replace the capability provided by the LAW80 weapon. The system is being developed in collaboration with Sweden, with Saab Bofors Dynamics as prime contractor and Thales Air Defence Ltd as the main UK subcontractor. The expected ISD is the end of 2006.

B6.13   The **Anti Structures Munition (ASM)** will provide the dismounted infantry with a hand-held, shoulder launched weapon to defeat defended structures such as buildings and bunkers. The programme is currently in a competitive Assessment Phase between Saab Bofors Dynamics and Dynamit Nobel Defence. Currently assumed ISD is around the end of the decade.

B6.14   A mid-life improvement (MLI) for the Shielder system, that lays the **L35A1 anti tank mines**, is scheduled for the middle of the next decade to extend the life to the middle of the following decade. The currently assumed **Barmine** OSD is early in the next decade, no replacement is planned.

B6.15   There is a huge variety of **explosives, pyrotechnics** and accessories, the detail of which is too diverse to cover. In addition, various capability gaps exist or will appear due to obsolescence, age, Taggant issues (plastic explosive tagging laws), IM compliance and other related issues.

Indicative planning assumptions

B6.16   Figure B6(i) illustrates our assumptions for spend in standard accruals terms on general munitions over the next 10 years. This includes the cost of new general munitions via the Equipment Plan (EP), the cost of maintaining in-service general munitions within the Short Term Plan (STP) and the planned research funding to identify and develop new technologies to support the EP. The UK is likely to spend around

£65M on general munitions next year across these three funding streams. The graph does not show spend on stock purchases which for FY04/05 amounted to approximately £208M. This is excluded from the graph in order to maintain consistency across all chapters. The total spend on general munitions is expected to remain broadly constant over the next ten years, and a general shift in balance from EP to STP spend as systems currently under development are delivered. We will continue to focus on developing new technical solutions, including IM technology, whilst maintaining and improving our existing capability.

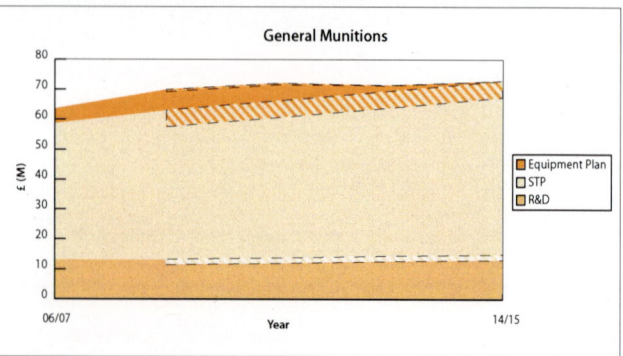

Figure B6(i) - Illustrative spend profile.
The above graph shows indicative spending in this sector over the next ten years. The figures from 08/09 are illustrative and include a range in order to emphasise the potential for shifts in investment priorities after the end of the current Spending Review period. This is prudent planning which does not distort the overall illustrative picture of general trends.

Munitions Vision – To achieve a defined military capability, when and where it is needed, at minimum whole-life cost, whilst ensuring intrinsic safety

What is required for retention within the UK industrial base?

B6.17   In order to deliver the munitions vision and to determine what should be retained onshore, the key is to ensure a sustainable and secure supply that provides best value for money.

System design

B6.18   It is essential that we retain onshore the Design Authority (DA) role and its underpinning capability for munitions manufactured in the UK. From a national security perspective, dependency on another nation for this may affect our ability to follow a preferred strategy, due to the provider nation having different strategic goals. We also require the ability to develop munitions for specific purposes to match our doctrine where it is not reflected by other nations. We need to maintain onshore an intelligent customer capability for non-UK designed munitions.

B6.19   We are a proven world leader in the field of IM and related energetic materials such as Polymer Bonded Explosives (PBX) and Low Vulnerability (LOVA) propellants. The maintenance of these capabilities in the UK can provide us with the ability to influence in co-operative/collaborative procurement through technology sharing.

System development

B6.20   We require little system development capability to be retained onshore in the UK. However, it is important that a UK-based platform system integration and system interface capability exists in order to ensure safe operation of any UK produced munition with the delivery system.

System manufacture

B6.21   We intend to reduce dependence on large munitions stockpiles through increased surge production, where technically possible and operationally viable and providing demonstrable value for money. As such, we wish to retain a substantive and flexible fill, assemble and pack capability onshore as well as a specialist steels and forging capability subject to this proving value for money. However, we do not consider it necessary to retain all aspects of bulk explosives manufacture in UK, though we want to retain PBX manufacture and casting. The position on SAA manufacture is less clear cut. It is desirable to retain SAA manufacture onshore, but not at any cost. We buy in the region of 120M rounds of SAA per annum. If offshore supply could be guaranteed there would be no need for onshore manufacture, but this would constrain the scope for supply chain compression and surge manufacture of a high volume product.

B6.22   In order to surge manufacture in general, and from a reduced onshore presence in particular, industry must ensure strong supply chain management. Furthermore, the surge manufacturing capability must be sufficiently robust to assure concurrent surge production across the required range of munitions.

Maintaining system capability through-life

B6.23   A robust through-life management capability onshore is vital. This includes surge production, munitions technical management, storage capability and intrinsic safety, duty of care and legal compliance activities. It is also essential that we retain a proof and surveillance capability onshore for UK designed munitions as well as at least a minimum munitions disposals capability.

Test and evaluation

B6.24   A UK-based Integrated Test and Evaluation capability is essential for quality assurance and some safety and operational security needs[1]. We do not consider it essential to carry out the testing onshore, yet it is vital that we retain the capability to understand, interpret and direct the testing to meet our performance and safety standards.

The 5.56mm ball ammunition is a good example of a piece of equipment manufactured by a UK Prime contractor (BAE Systems Land Systems) which integrates components from different suppliers:

Overview of the current UK defence market

B6.25   The general munitions marketplace is highly fragmented and cannot be termed a free market. The last 15 years has seen the pressure of globalisation, the collapse of the Far East market for munitions and the

---
[1] Currently we do not readily accept foreign test results where they do not meet UK standards.

effect of the post cold war peace dividend. This has resulted in the declining volume of requirements, falling R&D funding, considerable industrial consolidation and loss of domestic competition. For the main market players, choices have had to be made whether to exit the market, diversify and/or develop a long-term relationship with their nation state governments.

HMS ST ALBANS firing 4.5" Gun.

B6.26   Of particular importance to us is the capacity of the global market for security of supply. We see this most with the current global drain on the SAA market. The world market demand for SAA is estimated to be in the order of 5.2 billion rounds per annum. Following a period of decline since the mid 1990s consumption is rising, largely driven by the USA and also, to a lesser extent, by the UK. To fulfil its requirements the US is believed to be adding to its core SAA production, in order to retain surge capacity under US control. This suggests that US is not prepared to take the ultimate supply risk to operations; the UK has similar concerns.

B6.27   Figure B6(ii) shows the top 13 suppliers of general munitions, and support to general munitions during FY 04/05, a simplified breakdown of the products and support services provided by these suppliers is shown in the accompanying table. As can be seen, BAE Systems Land Systems, under the terms of the **Framework Partnering Agreement (FPA)**, supply the majority of our repeat buys of existing general munitions, achieving approximately 80% of the total value in FY04/05.

B6.28   The remaining 20% is subject to a healthy competitive environment. As such there are no grounds for immediate concern. However, the market is dynamic, and some companies are exploring acquisition or merger opportunities which could reduce the amount of competition. We will keep the situation under review.

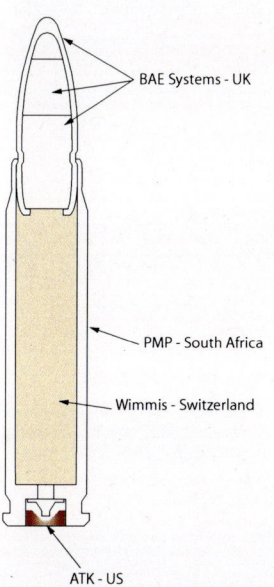

Headlines

80% of total spend with largest supplier BAE Land Systems

92% of remaining spend with 12 suppliers

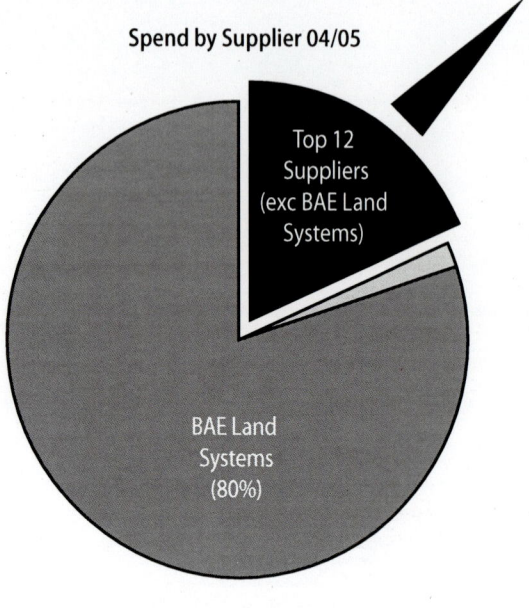

Spend by Supplier 04/05

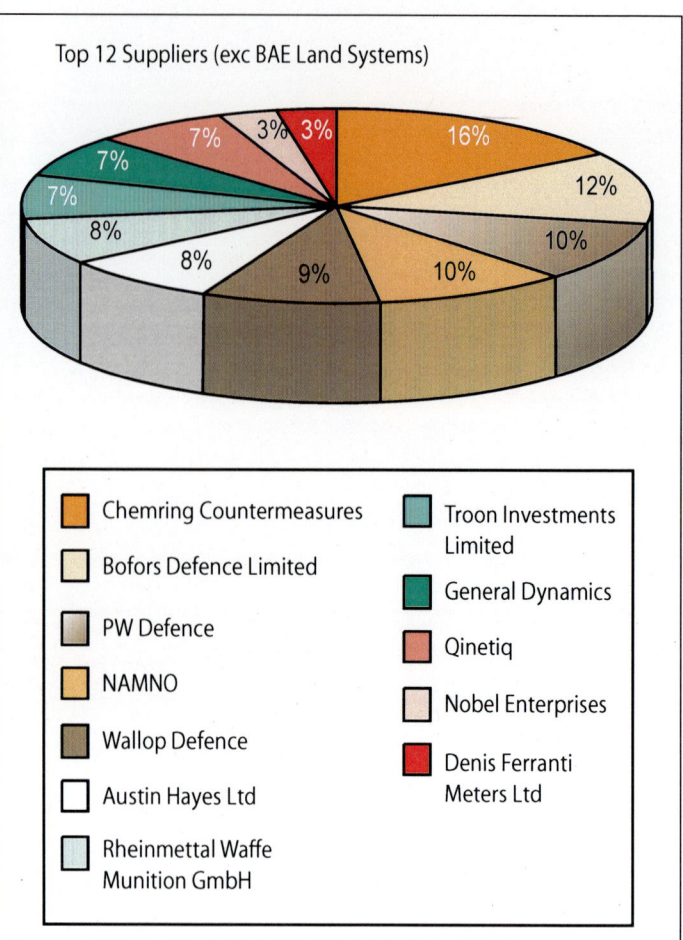

Top 12 Suppliers (exc BAE Land Systems)

Legend:
- Chemring Countermeasures
- Bofors Defence Limited
- PW Defence
- NAMNO
- Wallop Defence
- Austin Hayes Ltd
- Rheinmettal Waffe Munition GmbH
- Troon Investments Limited
- General Dynamics
- Qinetiq
- Nobel Enterprises
- Denis Ferranti Meters Ltd

Supplier	Product Range
BAE Systems Land Systems	Large Calibre Training Ammunition, Extended Range Bomblet System, Naval 4.5" Gun Prod HE and SUP, 120 Tank CHARM 3 Training Round, 81mm Mortar, 30mm Aden TP, 30mm DSRR Training Round, KCB/KAA Naval Rounds, Small Arms Ammunition, BLADE, DU Demil, Munitions Global Post Design Services, Naval Proof Yard, SX2 Explosive
Chemring Countermeasures	Aircraft counter-measures
Bofors Defence Ltd	105mm Illuminating
PW Defence	General pyrotechnics including: Smokes, Illuminating, EOD stores
NAMMO	66mm Anti-tank rocket
Wallop Defence	Aircraft counter-measures
Austin Hayes Ltd	Packaging
Rheinmetall Waffe Munition	General pyrotechnics, including smokes
Troon Investments Ltd	Mines & explosives
General Dynamics	Phalanx ammunition
QinetiQ	Technical support services
Nobel Enterprises	Nobel Enterprises
Denis Ferranti Meters Ltd	Smokes and Marine Marker pyrotechnics

Figure B6(ii).

B6.29   An example of healthy competition is the recent contract for supply of the Rocket Hand Fired Para Illum between PW Defence (UK), Rheinmetall (Germany) and others (who did not meet the User Statement of Requirement). Both PW Defence and Rheinmetall products were tested and the Rhinemetall product was chosen.

B6.30   Currently we retain in-house the through-life management of the munitions stockpile, including the management of disposal contracts. Storage, stock maintenance and distribution of munitions are also carried out by our personnel.

Sustainment strategy

The present

B6.31   The FPA was signed in 1999 and is a long term arrangement between us and RO Defence (now part of BAE Systems Land Systems). Under the terms of the agreement we will procure repeat buys of general munitions from BAE Systems Land Systems until 2010, on the assumption that the arrangement continues to represent best for value for money. The FPA has proved successful and continues to provide benefits for both parties.

B6.32   The FPA ensures that the UK maintains a strategically important sustainable and secure onshore industrial capability to deliver much of our general munitions requirement. It has been successful at achieving a long term value for money agreement. It secured products at fixed prices, resulting in a significant return on our initial investment and minimises the cost and risk of re-supply.

B6.33   The introduction of the FPA rationalised BAE Systems Land Systems munitions business and avoided the costs of any closures. Further, the formal alignment with us has supported investment in new technologies and products such as IM, whilst the formal adoption of gainshare mechanisms has created a culture of continuous improvement and provided the opportunity to increase profit rates.

B6.34   However, there are still recognised weaknesses to the current FPA as it does not adequately incentivise BAE Systems Land Systems to reduce its cost base and encourages the manufacture of product rather the provision of service. Significant effort has therefore gone into exploring alternative methods to improve the current arrangements, whilst continuing to emphasise the need for security of supply and value for money.

Partnering principles

B6.35   As a result of a capability analysis and building on the existing FPA, a partnering principles document was signed between us and BAE Systems Land Systems on 28 September 2005. The aim of the partnering principles was to reaffirm the strategic partnering intent in order to ensure long term security of supply, and incentivise BAE Systems Land Systems capital investment and rationalisation.

B6.36   Following a strategic review, and in line with the partnering principles, BAE Systems Land Systems therefore decided to reconfigure their general munitions business in the UK. This has resulted in plans to drive greater efficiencies in the BAE Systems Land Systems cost base whilst reinvesting to provide the capability in a more cost effective manner; this was announced publicly in October 2005. These plans are intended to provide secure, value for money supply of general munitions.

Relationship with other suppliers

B6.37   About 20% of general munitions are provided by other suppliers. These may be niche capabilities, low volume products, or simply provide better value for money. There are 75 extant contracts to a value of £200M.

B6.38   In the current financial year our largest non-BAE Systems supplier is PW Defence who provide a variety of pyrotechnic materials. We have a formal partnering agreement with the company based on the principles of the FPA with BAE Systems Land Systems. The largest non-UK based supplier is Rheinmetall, and we also have a formal partnering agreement with them. Partnering has encouraged strong working relationships with both companies.

B6.39   We have also created innovative contracting arrangements with the major suppliers of air countermeasures, Wallop Defence Systems and Chemring Countermeasures. These are designed to provide capability, and both companies are committed to providing a significant number of stores at 24 hours notice to meet surge requirements. The Wallops Defence Systems agreement has already matured into a formal partnering agreement.

B6.40   Other suppliers provide niche capabilities and specialist natures for a variety of roles and customers.

The future

B6.41   **Project MASS** (Munitions Acquisition – the Supply Solution), currently in its Assessment Phase, is a major MOD project intended to build upon the existing FPA with a view to securing, for the long term, an agreement that delivers best value for money to the taxpayer, and a sustainable way forward for industry.

MASS is expected to emulate the success of the FPA and maintain a key strategic onshore industrial capability.

B6.42   MASS is addressing all elements of the supply chain. It will include analysis of the potential benefits of industry delivering munitions directly to training areas, thus reducing the supply chain buffer stock and the associated cost footprint. It is also considering the case for reducing war stockpiles, replacing these with a wider surge capability, though we will not make any hasty judgements about moves in this area unless the operational implications are well understood and acceptable. Any changes which have implications for our staff will be subject to consultation with the Trade Unions in the usual way.

B6.43   MASS also seeks to address the perceived weaknesses of the FPA and to better incentivise our industrial partner to rationalise and reduce its cost base, whilst also addressing the potential of moving further towards capability based contracting via the vertical integration of the supply chain. Importantly, any MASS solution would underpin security of supply for general munitions, and value for money will be a key determinant. The options being assessed include non-BAE Systems Land Systems supply options.

Members of A Company Mortar Platoon, 1st Battalion, The Parachute Regiment, embed their 81mm mortars into position as they prepare a defensive position outside Basra.

B6.44   The 20% of items not provided by BAE Systems Land Systems may fall within the scope of MASS, depending on which option demonstrates the best value for money and maintains our general munitions capability. If the MASS option selected does not cover these items, then we are likely to investigate further enhancements to our other partnering agreements by widening their scope, and continuing to seek innovative contracting arrangements.

The way ahead

B6.45   We need to take forward Project MASS, with a view to making **decisions on how best to sustain our required access to general munitions in the summer of next year**, building on the joint working arrangement enshrined in the existing FPA as reinforces by the recently agreed MOD/BAE Systems Land Systems partnering principles. We are also actively pursuing partnering arrangements with other suppliers.

Definition

B7.1   Complex weapons are defined as strategic and tactical weapons reliant upon guidance systems to achieve precision effects. Tactical complex weapons fall largely into five categories: Air-to-Air; Air Defence; Air to Surface, Anti-Ship/Submarine (including Torpedoes); and Surface to Surface. In addition to conventional weapons the chapter also takes account of precision effects delivered by directed energy.

Vertically lauched Seawolf missile being fired from a Royal Navy Type 23 Frigate.

Strategic overview

B7.2   Complex weapons provide the UK Armed Forces with battle winning precision effects, which are able to achieve military advantage at a reduced level of asset use. The UK has over the past 10 years made a significant investment in the upgrade and development of complex weapons for the Armed Forces. This investment will peak at just over £1 billion next year for the delivery, research and support for complex weapons, and has allowed the UK both to develop a world class industrial capability, attract investment from the US and Europe and to deploy a relatively small number of highly capable high-value assets to meet military objectives. These assets need to be survivable and capable of extreme accuracy, ensuring that targets are effectively neutralised with the minimum of collateral damage and at the lowest risk to our own people. In order to maintain our military advantage the UK needs to exploit further complex cutting edge technologies such as directed energy to ensure that we have the capability to evolve our weapons to meet new and increasingly irregular threats. The threat environment does not stop adapting, nor should we. In addition continued complex weapons effectiveness and flexibility will also be dependent upon thorough integration into the emerging Network Enabled Capability (NEC).

B7.3   Complex weapons systems also underpin the maintenance of the UK's strategic deterrent, currently provided by the Trident missile system deployed on the Vanguard-class submarines.

During the 1991 Gulf War the percentage of Precision Guided Weapons dropped by the RAF was 20%; in Operation TELIC that balance was reversed, with 85% of weapons dropped being Precision Guided

Equipment programmes

B7.4   We have a large number of different complex weapons. Traditionally, these weapons have relied upon extensive maintenance and storage facilities to ensure their peak effectiveness. Weapons are now increasingly being designed to require little or no maintenance.   The requirement for complex weapons processing[1] will, however, remain integral to the delivery of operational capability, not least due to the large quantities of legacy weapons that will remain in service for some years to come.

Anti-Ship and/or anti-Submarine

B7.5   **Harpoon** is currently the Royal Navy's (RN) only long range anti-ship missile fitted to both Type 23 and Type 22 Frigates. It will require an upgrade to give increased capability against a wider irregular target set. The assumed Out-of-service-date (OSD) is well into the 2020s.

B7.6   **Sting Ray** is a lightweight anti-submarine torpedo in service and carried by RN (Helicopters and T23 Frigates) and the RAF (Nimrod). Stingray is currently undergoing an upgrade; **Sting Ray Mod 1** which is expected to enter service next year. Its OSD is assumed to be until the end of the third decade.

B7.7   **Spearfish** is a wire guided heavyweight torpedo and is the only anti-submarine and anti-shipping armament of the RN's submarine service. An replacement **(Submarine Launched Underwater Weapon, SLUW)** to this capability is being planned, including Insensitive Munition (IM) compliance, either based on Spearfish or another military-off-the-shelf (MOTS) weapon, and is assumed to enter service early in the next decade.

Air Defence (including maritime)

B7.8   **Sea Dart** provides area maritime air defence capability and is launched from T42 Destroyers; it will progressively be replaced by **Principal Anti Air Missile System (PAAMS)**, firing Aster 30 and Aster 15 missiles, as the RN's T45 Destroyers enter service towards the end of the decade.

B7.9   **Seawolf** is the RN's point defence missile system fitted to all frigates for hard-kill defence against aircraft and anti-ship missiles.

---

[1] *Processing refers to any weapon related activity acting upon either energetic or inert components (e.g. assembly, testing) other than operations needed in support of storage and transportation.*

The Block 2 follow-on missile run has recently entered service and is envisaged to be in service until the end of the third decade.

26 SQN RAF Regiment launch a rapier missile during Exercise Saif Sarea II in Oman.

B7.10   **Rapier Field Standard C (FSC)** provides low level air defence over the battlefield and its planned OSD is at least 2020. **High Velocity Missile (HVM)** System, commercially known as **Starstreak**, is a very short range air defence weapon designed to attack helicopters and low-flying aircraft. Its OSD is planned to be well into the 2020s.

Air-to-Air

B7.11   The RN and RAF's short-range missile capability is provided by **AIM-9 (Sidewinder)** but has recently been enhanced by the introduction of the **Advanced Short-Range Air-to-Air Missile (ASRAAM)**, next generation short-range Air-to-Air Missile that entered full service in 2004. The anticipated OSD for ASRAAM is well into the 2030s.

B7.12   We have two variants of the **Advanced Medium-Range Air-to-Air Missile (AMRAAM)**:

- **AIM120B** has been in service since 1995 and is currently carried on Sea Harrier and Tornado F3 and will be on Typhoon. Expected OSD is for the middle of the next decade.

- **AIM120C-5** is the current production variant of the AIM120 family and is being procured by the UK as an interim weapon pending the introduction of Meteor for Typhoon. It will enter service in 2006 and its expected OSD is in the middle of the next decade.

B7.13   **Beyond Visual Range Air to Air Missile (BVRAAM)** also known as Meteor is a collaborative programme between the UK and five other partner nations currently planned to enter service early in the next decade.

Air-to-Surface

B7.14   **Air Launched Anti Radiation Missile (ALARM)**, is a medium-range, anti-radar, Air-to-Surface missile integrated into the Tornado GR Mk 4. Its OSD is planned for early in the next decade. ALARM forms part of the Suppression of Enemy Air Defence (SEAD) capability. A variety of soft-kill options are currently being explored in relation to the capability currently provided by ALARM.

B7.15   **Hellfire** is a Radio Frequency (RF) or laser guided Air-to-Surface missile for the AH64 – Apache helicopter and is expected to be in service until the beginning of the third decade.

B7.16   **Maverick** anti-armour missiles were procured as an Urgent Operational Requirement (UOR) for use by Harrier GR7 at the start of Operation TELIC and are expected to remain in service until the middle of the next decade.

B7.17   **Storm Shadow** is a long range missile with a capability against hardened targets. The missile entered service in 2004 and its performance characteristics are planned to be progressively enhanced through an incremental technology insertion programme. The weapon is expected to continue in service beyond 2030.

Storm Shadow – Conventionally Armed Stand-Off Missile (CASOM).

B7

B7.18   Paveway II (Laser) and **Enhanced Paveway II** (Laser/GPS) guidance kits are being procured to provide guidance capability for the UK's 1,000 lb bomb. Paveway III (Laser) and **Enhanced Paveway III** (Laser/GPS) bombs will remain in service beyond 2015. **Paveway IV** (Laser/GPS) bombs will enter service around 2007 and are anticipated to become the mainstay of the RAF's Air to Ground bombing capability for at least 20 years.

B7.19   **Brimstone** is an advanced air-launched anti-armour weapon that entered service on Tornado GR4 in 2005. Brimstone will be integrated onto Harrier GR9, and is a candidate weapon for Typhoon. It will remain in-service until well into the 2020s.

B7.20   **Selected Precision Effects at Range (SPEAR)** is planned to address the capability requirement to be able to attack fast-moving targets at range. This programme is currently in its concept phase.

B7.21   **Future Anti-Surface (Guided Weapon) FAS(GW)** is planned to provide capability to attack fixed coastal targets and highly manoeuvrable vessels.

Surface-to-Surface

B7.22   **Guided Multi Launch Rocket System (GMLRS)** is an IM-compliant unitary missile. To ensure economies of scale and logistic support benefits, we have chosen to co-ordinate with US production (ratified by a bilateral Memorandum of Understanding), its expected ISD is 2007, and is assumed to remain in-service until early in the second decade with M270 and well into the 2030s with Lightweight Mobile Artillery Weapon System (Rocket) (LIMAWS(R)).

B7.23   **Indirect Fire Precision Attack (IFPA)**, this programme is intended to address the requirement to be able to attack static, manoeuvring and mobile targets with precision. The capability is expected to be delivered by an initial operating capability based largely on existing technology towards the end of the decade, with incremental enhancements over its service life.

Complex Weapons

B7.24 The **Light Forces Anti-Tank Guided Weapon (LF ATGW)** is a man-portable, medium range guided anti-armour missile system, which will provide the Infantry and Formation Reconnaissance with the ability to defeat modern and emerging armour threats. This requirement is being met by the US Javelin system and replaces the Milan and Swingfire missiles. ISD was achieved in July 2005 and the Weapon is expected to remain in-service well into the 2020s.

Javelin is launched from a Pinzgauer vehicle by 42 Commando Royal Marines during demonstration held at Imber Clump, Warminster.

B7.25 Tomahawk Land Attack Missile (TLAM) provides a land attack capability against high value, non-hardened facilities in heavily defended areas. The new Block IV missiles will bring additional capability over the current Block III standard missile. They will be carried by our attack submarine fleet and are expected to enter service in 2007 with an approximate OSD of at least 2040.

Indicative planning assumptions

B7.26 We will, on current predictions, spend over £1 billion in the Equipment Plan on new complex weapons in FY2006/2007. The scale of our investment in new systems is at a peak and will reduce by some 40% over the next five years as production activity begins to decline following the delivery of systems such as Storm Shadow and Brimstone.

B7.27 Given the essentially incremental nature of our planned approach to future capability development, there is, apart from the Meteor programme, little significant planned design and development work beyond the next two years. This will present a substantial challenge as we seek to maintain those industrial capabilities we would wish to retain on-shore.

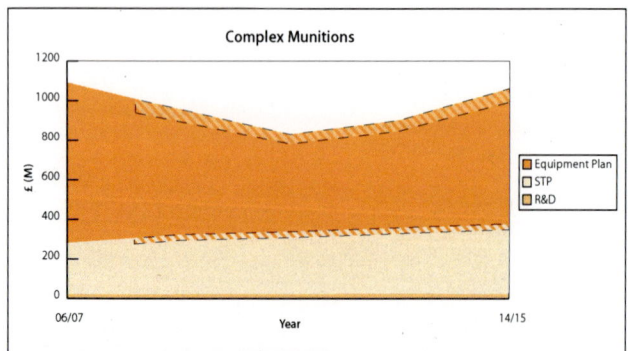

Figure B7(i) - Illustrative spend profile.
The above graph shows indicative spending in this sector over the next ten years. The figures from 08/09 are illustrative and include a range in order to emphasise the potential for shifts in investment priorities after the end of the current Spending Review period. This is prudent planning which does not distort the overall illustrative picture of general trends.

What is required for retention in the UK industrial base?

B7.28 Given the significance of some types of complex weapons, in many cases we require absolute confidence in the performance and safety aspects of our weapon systems, particularly as they become more sophisticated. This can only be guaranteed if the UK has access to and a comprehensive understanding of, the entire system and its design and controlling software. Full access to this mission critical software and information cannot always be secured when procuring complex weapons from offshore suppliers. While we have procured from offshore in the past, we have balanced this risk against the military capability afforded by the overseas options and also against the assumption that we could continue to rely on a sustainable UK industrial base. This does not preclude future procurements off the shelf or otherwise from other nations, even where our access to critical software and information is denied. But it means recognising the risks of such approaches, and retaining the option to avoid these through a sovereign capability onshore, particularly for campaign-critical capabilities where there is little competition other than from US companies, is highly desirable.

B7.29 Therefore to maintain appropriate sovereignty, it is important that the UK can use, maintain and upgrade specific capabilities in its inventory, independent of other nations. To do this, the UK needs guaranteed access to the following key functions:

- Provide weapon systems design and performance expertise independently or as a leading player in collaboration with other nations.
- Understand threats from technology proliferation.
- Exploit emerging and novel technology (outlined in detail in the table at the end of the chapter).
- Develop Counter Counter-Measures (CCM).
- Respond to UORs.
- Undertake national projects within the relevant legal and international frameworks.
- Retain sufficient understanding of elements of the nuclear deterrent[2].
- Through-life capability management and support for the current weapon inventory.

B7.30 In order to deliver these key functions we need access to a UK industrial base that, as a minimum, provides the following capabilities:

Concepts generation, design and systems engineering

B7.31 In order for the UK to act independently it will be critical to retain on-shore knowledge and experience of how the weapon system can achieve its objective against an adversary in a hostile environment. This means we need a critical mass of expertise to develop complex weapons concepts through synthetic environments and the ability to design and integrate the weapon with platforms and sensors.

Exploiting and controlling complex weapons in a network environment

B7.32 Increasingly, complex weapons will need to be networked[3] with other systems to fully exploit their potential and to be an information provider to the network. This will require the ability to modify the mission critical software

---

[2] A number of UK suppliers are involved in modelling and assessment of the current Deterrent and would be involved in any studies for any potential replacement.
[3] For example, the information network, mission planning, command and control, targeting sources, weapon dynamics, propulsion and lethal package.

and to manage weapon and network interfaces. The capability to design the actual guidance, control, and seeker algorithms of the weapon is therefore vital to operational success. This capability is central to delivering rapid, flexible and precise military effects, reducing collateral damage, maintaining the existing inventory effectiveness and reducing our vulnerability to counter-measures. The UK's ability to design, understand and/or have access to such critical information will be a key measure of sovereignty.

Underwater capability - Torpedoes

The UK must retain the capability to support the current inventory and also to write tactical software, (fusing, guidance and control algorithms) and to design and integrate homing heads. Provided we maintain control of these elements, we would be prepared to buy new torpedoes designed and manufactured overseas.

Lethal package and propulsion

B7.33   These are the key determinants of the weapon systems performance. These elements by their nature carry a significant safety risk that requires a range of specific capabilities if the UK is to be a safe owner. To improve safe ownership; all of our new weapons will need to be IM-compliant.

B7.34   The design of the lethal and non lethal package is required to ensure proportional effects including low collateral damage. This leads to a requirement for expert knowledge of the conventional and novel warhead and the ability to design the safety and arming functions to be retained in the UK industrial base.

Directed Energy Weapons (DEW)

We judge that Directed Energy technologies (lasers and radio frequencies) could be highly significant in the future, particularly for protecting our Forces from a range of threats, including Improvised Explosive Devices. They could also offer the UK non-kinetic and / or less-lethal options to replace, enhance or complement traditional kinetic weapons, such as missiles, and offer significant opportunities to reduce collateral damage, notably in urban areas. Due to the reusable nature of the technology and some of its potential applications, there is also potential for a significant reduction in the logistic footprint and whole life costs compared with conventional weapons. We are assessing the potential military utility of DEW technology through a number of research programmes. Aspects of these programmes include technology development, user requirements and concepts of operations.

B7.35   We will need to support and re-life as necessary in-service rocket motors and warheads. Some specific types of propulsion, energetics technology and manufacture are unique to the UK and in particular the UK has the lead in IM which we will need to retain for the future safe ownership of our weapons.

Through life support/technology insertion

B7.36   The need to sustain the current inventory, have sovereign control over weapon deployment and the ability to upgrade, at least for part of the inventory, requires the retention of an onshore ability to maintain complex weapons through design authority and through-life support processes. This is often referred to as complex weapon processing.

Through life support to sidewinder © Ultra

Overview of the global defence market

B7.37   The world missile market is dominated by two principal US primes; Raytheon and Lockheed Martin. The largest European prime is MBDA, a joint venture owned by BAE Systems, EADS and Finmeccanica. Within Europe, MBDA has national representation in the UK, France and Italy as well as a presence in Germany. The European sector is also characterised by a series of smaller, contractors such as Thales and Saab Bofors Dynamics that also have the capability to act as a prime contractor.

Overview of the UK defence market

B7.38   In the UK the main player in the missiles sector is **MBDA UK Ltd**, the prime contractor for around half of our in-service inventory and account for over 50% of the investment currently on contract.

B7.39   **Thales Air Defence Ltd (TADL)** is the prime for High Velocity Missile (HVM) and has had a niche capability in Short Range Air Defence as well as Electro Optic Counter Measures (EOCM).

B7.40   **Raytheon Systems Ltd** is a wholly owned UK subsidiary of the Raytheon Company USA. It is the prime contractor for the Precision Guided Bomb (Paveway IV) and is also a major sub-contractor with particular capability in missile electronics, where it is an exporter to US programmes such as AMRAAM, Tomahawk, HARM and TOW.

B7.41   **Roxel UK Ltd** is the UK's only designer and manufacturer of rocket motors and accounts for some 60% of our requirements for rocket motors.

B7.42   **BAE Systems Underwater Systems Ltd** is the single UK indigenous company capable of designing, developing, supplying and supporting in-service light and heavyweight torpedoes.

B7.43   Other companies in the UK with elements of the complex weapons sub-system capability include **Thales Missile Electronics Ltd** (seekers and fuzing), **BAE Systems Land Systems Ltd** (warheads), **Selex UK Ltd** (seekers), **Insyte** (trackers), **LM (UK) Insys Ltd** and **QinetiQ**, (intelligent customer advice, research, design, in-service support, test and evaluation). These companies have had success supplying to the wider international missile market.

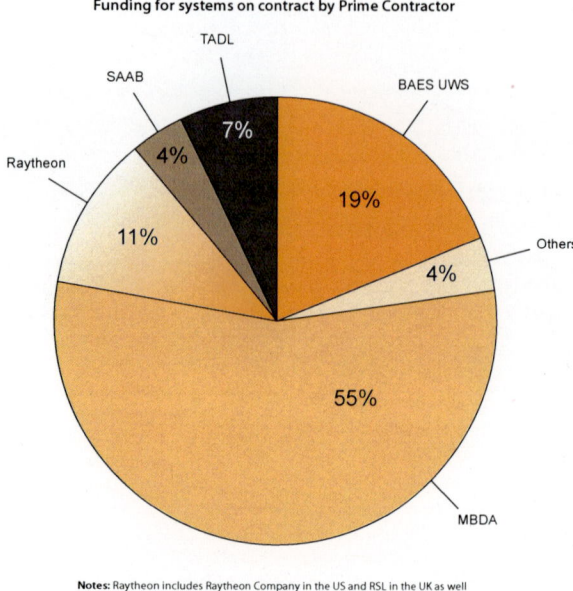

Funding for systems on contract by Prime Contractor

Notes: Raytheon includes Raytheon Company in the US and RSL in the UK as well as Raytheon/Lockheed Martin Joint Venture. On Contract includes all funding once Demonstration phase contract has been placed, system may not yet be in-service.

Figure B7(ii).

Sustainment strategy

B7.44   The UK needs to retain the capability within industry to design, develop, assemble, support and upgrade complex weapons. Future demand and investment will not support the UK's current spectrum of industrial capability in the way it is currently provided from 2007 onwards.

B7.45   The fragility of the wider UK industrial base is such that unmitigated open international competition will put the sustainment of key industrial capabilities at risk. We need to consider how we ensure that our requirements can be met, by a sustainable industry in a value for money fashion into the future. We intend to work with all elements of the onshore industry, therefore, over the next six to twelve months to establish whether – and if so how – this can be achieved. This dialogue will need to be based on the following key principles:

● **A tempering of competition:** For the short to medium term, we will consider suspending the use of international competition to meet our future complex weapons requirements (with the exception of torpedoes). We will instead be looking at whether, if in this period our requirements were to be directed to onshore industry, it would be possible to meet them, secure long-term value for money, and maintain a viable industrial base. This would of course be subject to the construction of a satisfactory business case and appropriate contractual arrangements. However, even more fundamentally, it is unlikely that the resources currently available could sustain the industry in its current form, and industry must play its part in meeting the challenge.

● **Industrial restructuring:** Given the over- capacity in the domestic and wider European market and the trans-national nature of the predominant onshore industrial player, there is significant potential for industrial rationalisation and consolidation. We will, as part of the work mentioned above, need to work closely with our European partners to identify whether a coordinated approach to sustain a viable industrial base is possible. But this will not be to the exclusion of other US-owned companies, in particular those who have already established a firm foothold, and, consistent with our definition of the UK defence industry, are focussed on securing appropriate technology access in the UK for us to maintain appropriate sovereignty. This will continue to be welcome, but it will be important to ensure

that we do not sustain overcapacity in the onshore industrial base or act in a way that, overall, weakens the sustainability of the capabilities and technologies to which we attach importance.

● **Different approaches to acquisition through-life:** We need to explore different approaches to acquisition and support so the required level of capability can be maintained. In addition to restructuring the industry, including the supply chain, this will require:

  ● an open dialogue between all parties;
  ● working to improve arrangements for through-life capability support;
  ● agreed mechanisms to demonstrate long-term value for money.

B7.46   Changing the way we do business will be challenging, for both us and industry, given the number of companies in the sector set against the declining workload expected on new developmental programmes.

B7.47   We also need to take into account that the UK has overcapacity and capability when it comes to managing the maintenance, support and storage of complex weapons. Much of this capability resides within the Defence Logistics Organisation (DLO) Defence Munitions Centres (DMC), which should be retained in Government ownership. It is clear that greater integration and rationalisation of the industry and our support capabilities could be achieved. Close MOD and industry support relationships are also likely to enable easier technology insertion and spiral development. Future partnering and commercial arrangements for support are therefore being explored at both the individual project and the strategic levels; we will be considering whether Contracting for Availability could bring benefits to both MOD and industry. This will be taken into consideration during the joint work with industry over the next six to twelve months. Any changes to our current arrangements arising from this work which impact on MOD staff would be subject to consultation with the Trade Unions in the usual way.

B7.48   Our approach to torpedoes will largely be centred on the outcome of the SLUW programme. We will need to consider carefully how to sustain our world-class capability in torpedo homing head design and integration if an overseas option is chosen to provide this upgrade. If this situation arose, we would need to consider bringing the capability into our research facilities to ensure continued secure access.

B7.49   **Research and Development:** Research and Development is important to the maintenance of cutting edge technology. We will need to develop a series of route maps to link technology exploration to future capability and exploit the high value intellectual capital built up through years of investment. This will need to be co-ordinated across the complex weapons sector to remove duplication. As part of our considerations of the best means to sustain key capabilities and technologies, we will need to be aware of the role that targeted Technology Demonstrator Programmes might play.

B7.50   **Looking to the future:** There are obvious attractions – given the through-life costs of supporting the large number of different weapons in current and planned service – in examining with industry how we might rationalise our inventory. In addition to the associated cost savings, this might allow reductions in the overall stockpile, extend the time over which production takes place (which could allow easier technology insertion and smoother production throughput) and increase our flexibility in operations. This will require us to work closely with industry to develop technology roadmaps and maximise the benefits of common sub-systems technology.

B7.51   The challenges in this sector are very real and the approach we have laid out will be very demanding. We will need to work fast, in conjunction with industry if we are to avoid seeing the UK industrial capability going into decline. A joint MOD and UK industrial team will be created in the immediate future to take forward this work.

The way ahead

B7.52   We need to establish a multi-disciplinary team charged **with working with the onshore industry to establish how we might together seek to sustain the critical guided weapons technologies and through-life support capabilities that we judge to be so important to our operational sovereignty.** Given the trans-national nature of the industrial players, this dialogue will need also to engage our allies and partners, particularly in Europe. This work will be complex and will necessarily take time, but our intention is that we should have a clearer way ahead by the end of 2006, with a view to informing decisions in 2007.

B7

Complex Weapons

Definition

B8.1   C4ISTAR (Command, Control, Communication and Computers, Intelligence, Surveillance, Target Acquisition and Reconnaissance) comprises those capabilities, that when linked can achieve the desired military effect by providing the ability to command and inform the Armed Forces in an effective and coherent manner and that together form the initial stages of Network Enabled Capability (NEC). They generate and enable the delivery of accurate, timely and appropriate information and intelligence products to decision-makers at all levels in national and coalition operations. The capability area can be sub-divided into two components; Command, Control & Information Infrastructure (CCII) and ISTAR. The C4 component of CCII provides optimised and integrated Command and Battle Management (CBM) and Global Information Infrastructure equipment capabilities. The ISTAR component constitutes a combination of sensors, weapons systems, IT hardware & software, people and processes that collectively enable the ISTAR cycle (Direct, Collect, Process & Disseminate). The technologies associated with these capabilities are often leading edge, draw extensively on research activities, exploit developments in the civil sector, and place a premium on innovation, rapid technology insertion and effective system integration[1].   It is this sector that the principle of spiral development to enable continued operational performance is most relevant given the rapid pace of technological development.

Strategic overview

B8.2   The 2004 White Paper 'Delivering Security in a Changing World: Future Capabilities' stated that the continued transformation of UK forces is dependent on exploiting the benefit of NEC. It also noted that NEC, by enabling 'the ability to respond more quickly and precisely, will act as a force multiplier enabling our forces to achieve the desired effect through a smaller number of more capable assets.' NEC is one of our highest priorities for future investment in research, acquisition, people and training. The delivery strategy focuses on three Epochs: Initial State, achieved by 2009 and characterised by interconnection; Transitional State, achieved by the middle of the next decade and characterised by integration, and Mature State, with an aim of achievement by the middle of the second decade onwards and

---

[1] Given the large number of defence capabilities that have some C4ISTAR functionality as a second or lower order function of the system in which they reside e.g. platform mission and command systems, weapon locating and guidance radars; navigation and aircraft defensive aids, and in order to establish a manageable boundary for this Chapter, the above Definition of C4ISTAR has been interpreted in a relatively literal sense. Thus, in considering future defence capability demand and associated industrial and technological capabilities, the analysis concentrates on those systems and applications whose core function is to Command and Inform the Armed Forces as defined above. These are principally the defence capabilities for which requirements are set by the Equipment Customers Directorates CCII and ISTAR. In consequence, some of the industrial and technological capabilities identified will have relevance to defence capabilities discussed in other Chapters and vice versa. In a similar vein, business information system applications supporting functions such as defence logistics and personnel or medical services are addressed only superficially, on the basis that enabling technologies and industrial capabilities import principally from the civil commercial sector, the larger customer base for which will drive innovation and functionality that defence applications can exploit.

characterised by *synchronisation*. However, the pace of change is such that the delivery strategy must continue to evolve as further opportunities for NEC delivery present themselves in accordance with technological innovation.

B8.3   It will be the C4ISTAR related capabilities that will help underpin the overarching NEC capability by providing the technology to enable agile, networked and informed armed forces. There is thus an enduring requirement for some industrial and technological capabilities across the spectrum of acquisition from system design to maintenance of the capability, and for systems integration, system engineering and information assurance skills. In considering the industrial base from which future C4ISTAR capabilities can be derived, we must also look beyond our current providers to other sectors, from mobile phone providers to film studios, who may be able to offer innovative applications or derivations of non-military technologies in support of the desired capability.

High Level Network Map.

B8.4   The increasing incidence of asymmetric operations in difficult environments (including the urban) against irregular forces and individuals who are increasingly aware of our ISTAR capabilities is driving a requirement for increasingly detailed, unambiguous, persistent and timely information. At the same time, peace enforcement and peace keeping operations require persistent surveillance in a more benign environment.

B8.5   It is assumed that the longstanding strategic relationship with the US will continue for the foreseeable future. In particular, with respect to ISTAR, it is assumed that access to US research and industrial capabilities will continue on a similar footing, albeit we are operating with continued uncertainty regarding current and future access to US technology.

Equipment programmes

B8.6   There are a variety of programmes that are either in initial planning stages, are due to enter service soon or have recently entered service that provide capability over three principal areas; **Command and Battlespace Management**, the **Network** and **ISTAR**.

B8.7   **Command and Battlespace Management** are programmes that are envisaged to provide commanders with the data and information that they need.

B8.8   Platform Battlefield Information Systems Application (P-BISA), which is a part of the wider **ComBAT, Infrastructure and P-BISA (CIP)** programme, will provide hardware and software to enable armoured vehicle commanders to send and receive information over the Bowman tactical internet. It will provide secure voice and data messaging, situational awareness and operational planning tools. As part of the Bowman programme the contract for P-BISA was awarded to General Dynamics (UK). Just over 1000 systems will be delivered to equip Warrior, Challenger 2 and Combat Vehicle Reconnaissance Tracked (CVRT) Scimitar.

B8.9   Land Environment Air Picture Provision (LEAPP), is envisaged to provide a near-real time correlated air picture for the land environment world-wide, either alone or as an element of integrated air defence, to enable a graduated and enduring contribution to the NEC community in national and multi-national operations in a multi-threat environment. It is currently in its assessment phase.

B8.10   J2CSP is the **Joint Command and Control Support Programme**. It will develop a command support application hosted on the Defence Information Infrastructure (DII) that bring together three existing systems, (RNCSS, RAFCIS and JOCS) and extends functionality. This complex programme has four constituent parts all of which are in the Concept or Assessment phase.

B8.11   The Deployable Air Traffic Control Capability Enhancement (DATCCE) programme aims to increase our DATC capability to enable simultaneous support to an additional two bare Deployed Operating Base locations. It will allow air traffic controllers to provide a full instrument flight rules service, 24 hours a day in all weathers, to enable military aircraft to operate safely from a number of austere and bare base operating locations as part of JRRF operations.

B8.12   A hierarchy of **logistic** capabilities are being developed to deliver a Joint Logistic Picture to provide the necessary decision support to Logistic commanders at all levels in the deployed force, and asset tracking end-to-end throughout the Joint Support Chain. These include JC2SP, as well as 'transactional' systems: Management of the Joint Deployed Inventory (MJDI), Joint Asset Management and Engineering Solutions (JAMES) and Management of Materials in Transit (MMIT). The transactional systems will grow incrementally to encompass all defence assets and commodities:

- MJDI will provide inventory management and the auditable account for all material held in Defence outside the Base Depots.

- MMIT will provide a management information layer above asset tracking systems in order to manage supply chain operations.

- JAMES will deliver engineering and asset management of all equipment and key equipment components across defence, including the management information system to support Whole Fleet Management and the exploitation of HUMS capability delivered through platform projects.

B8.13   The Joint Military Air Traffic Services (JMATS) programme aims to provide the next generation of military terminal air traffic control services. It will support operations involving all three services at frontline airfields, training airfields, ranges and contracted airfields.

B8.14   As the framework nation for HQ Allied Rapid Reaction Corps, the UK is required to provide Command Intelligence Systems (CIS) for the headquarters on operations, part of which is the Command Support System. The **ARRC Command and Control Information System (ARRC C2IS)** will comprise applications (and associated infrastructure) that will support a flexible and integrated orders creation and dissemination process, situational awareness, planning, synchronisation and information management. Full delivery is envisaged in the latter half of this decade.

B8.15   The **UK Air Surveillance, Command and Control System (UKASCACS)** project is intended to investigate options to provide an integrated Air C2 system capable of planning, tasking and executing operational level through to tactical level, air battle management and surveillance in the UK.

B8.16   Future Integrated Soldier Technology (FIST) will enhance the ability of the dismounted infantry to move, find and engage the enemy. It will provide an integrated suite of capabilities adjusted for the soldier's role including improved protection, day and night surveillance and target acquisition and assistance with navigation, command and control and battle preparation. The programme is currently in an Assessment Phase which is supported by Thales, and is envisaged to come into service around the turn of the decade.

B8.17   Network programmes are designed to facilitate the rapid and secure communication of data:

B8.18   Bowman is providing the Armed Forces with a tactical communications system for all three services in support of land and littoral operations. In addition to secure voice communications there are a set of common software tools that will enhance situational awareness at all levels and aid planning for, and control of, operations. General Dynamics (UK) are delivering this project. Bowman is likely to remain in-service well into the 2020s.

Bowman is the UK's tactical communications system, providing data and voice communications and a satellite-precision navigation to the front line for all 3 services.

B8.19   Skynet 5 is delivering the next generation of military satellite communications and will, incrementally, replace the Skynet 4 ground and space segment. Skynet 5 is being procured as a Private Finance Initiative (PFI) and its boundary has been extended to include Maritime and Land terminals making the commercial partner, Paradigm, responsible for the end to end delivery of satellite communications services. Capability from Skynet 5 is already being made available for the Armed Forces and should have a service life of at least 20 years.

B8.20   Cormorant is a new capability, that has recently entered service, designed to meet the needs of the Joint Rapid Reaction Force (JRRF). It provides communications within and between the deployed Headquarters of the Joint Force. A high capacity, secure communications system, Cormorant will enable the Joint Task Force Headquarters to command all of the subordinate Joint Force Component Command Headquarters. The system can be mounted in either vehicles or transit cases to enable its modular and flexible use on expeditionary operations. The vehicles can fit onto the RAF's C-130 aircraft to enable rapid deployment. It will be fully interoperable with other UK communications systems such as Ptarmigan, Bowman and Skynet 5 and allows communications between UK and Allied headquarters. EADS UK Ltd is delivering this capability.

B8.21   The **Defence HF Communications Service (DHFCS)** provides a coherent solution to the defence strategic High Frequency (HF) communications requirement until the end of the decade. It is bringing the disparate maritime and air systems together under a single service provision that will also capture the HF facilities in the overseas commands. The service provider is responsible for both the technical solution and the provision of HF services and is expected to introduce new services and improved technology on an incremental basis throughout the programme. It is expected that the new service will deliver a step change in data throughput, improved reliability through the exploitation of new automated processes and a reduction in manpower required to operate the system. The DHFCS contractor is VT Communication.

B8.22   **General Key Management (GKM)** has recently entered service and is a tri-service project to replace the current manual distribution of un-encrypted, hard copy material, with a computer controlled system, which will manage and distribute encrypted key material stored on magnetic media. Currently all Communication Security material (key variable, cryptographic equipment and publications) is managed by the use of manpower intensive hand written records. A manual system will not be able to handle the predicted increase in key material. GKM will manage the existing physical material for some of our high grade cryptography products. The GKM contractor is EADS UK Ltd.

B8.23   The **RNJTIDS/STDL** project will provide a secure, Electro-Magnetic Capability resistant, high capacity, digital Tactical Data Link capability (Link 16), to the Royal Navy and is about to come into service. This capability will overcome the critical limitations of Link 11 for the exchange of near real time tactical information in the line of sight domain, and will maintain interoperability with Allies. The related STDL (Secure Tactical Data Link) programme (now merged into RNJTIDS), approved in the mid 1990s, extends use of the same L16 messages incorporated into command systems under the RNJTIDS programme, to Beyond Line Of Sight via UK Super High Frequency satellite bearers.

B8.24   **FALCON** is intended to provide a tactical formation level secure trunk communications system for the UK and the Allied Rapid Reaction Corps (ARRC). It is planned to replace the capability provided by Ptarmigan, RTTS/DLAN and EUROMUX. FALCON should contribute to the 'Resilient Information Infrastructure' theme of NEC by providing the modern, secure communications infrastructure required by deployed formations and operating bases, such as Bowman and Cormorant.

B8.25   **Computer Network Defence** is planned to provide an ongoing capability to protect our network above existing defensive measures. It will consist of an integrated set of mechanisms, processes and organisations, and an effective command structure, which together will maintain our capability to defend our Information Systems (IS) from attempts to disrupt or disable them.

B8.26   **Physical Infrastructure for Deployable Headquarters** will provide environmentally protected and appropriately deployable physical infrastructure for dismounted headquarters and communication nodes, in order to ensure that off the shelf C4I equipments do not fail in harsh environments. The capability will enter service around the turn of the decade.

B8.27   **The Defence Information Infrastructure (DII)** will provide a networked IS infrastructure, at all classifications and caveats of information, to all non-deployed units and headquarters and certain deployed elements. The programme is being delivered incrementally by the Atlas Consortium.

Satellite communications provide a rear link between overseas deployments in the UK.

B8.28   **The Defence Fixed Telecommunications System (DFTS)** provides our voice and data fixed telecommunications services. This is currently being delivered through a 15 year Public Private Partnership (PPP) deal signed with British Telecom in 1997 and extended until 2012 this year.

B8.29   Project **ANSON** is envisaged to provide defence users, outside the battlespace, with an efficient high grade messaging capability to exchange the correct information, with the correct people, at the correct time, in the correct format and in a legally accountable manner. It is planned to complement the medium grade messaging capability and will be delivered as part of DII, in the latter half of this decade.

B8.30   The **Joint Network Integration Body (JNIB)** is a joint MOD/Industry body empowered to intervene and deliver solutions to integration problems between component joint planning networks. JNIB is being delivered using a combination of existing MOD resources and directly contracted industry resource from the key companies concerned.

B8.31   In addition, we also invest in bodies such as the **Integration Authority (IA)** which works towards the integration of information and projects across the whole battlespace in order to enhance military capability in the conduct of joint and combined operations.

B8.32   **ISTAR** programmes planned to facilitate the processes needed to acquire and analyse data for the Armed Forces:

Sentinel R Mk 1 at take-off.

B8.33 The Airborne Stand Off Radar system (**ASTOR**) will give our Armed Forces an entirely new capability, providing commanders with accurate and timely ground surveillance information in a wide range of scenarios, from humanitarian aid to combat operations. As a key element in our future mix of sensor assets, ASTOR will play a significant role in our efforts to develop NEC. ASTOR is being developed by Raytheon Systems Ltd. The five ASTOR air platforms, based on a modified Bombardier Global Express aircraft, will be known as the Sentinel R Mk 1. They are currently expected to begin to enter service from the end of 2006.

B8.34 **SOOTHSAYER** incorporates some of the world's most advanced electronic warfare (EW) equipment. The system will comprise both electronic support measures and electronic counter measures (ECM). It will be fitted to high mobility light role and armoured vehicles and will be fully integrated in to the digitised land battlespace and be interoperable with other EW systems in joint operations. Lockheed Martin Systems Integration was selected as prime contractor in 2003, following a three year competitive assessment phase. SOOTHSAYER equipment will start to enter service towards the end of the decade.

B8.35 **SHAMAN** intends to use Commercial Off the Shelf (COTS), open architecture systems to provide the capability for communication electronic support measures to Royal Navy warships. A UK/US industry team led by BAE Systems Insyte was selected in early 2005 as the preferred contractor for an Advanced Demonstration phase for the project, aiming to mature and de-risk the SHAMAN solution

B8.36 The **WATCHKEEPER** tactical Uninhabited Aerial Vehicle (UAV) system will provide UK commanders with accurate, timely and high quality information, including imagery. WATCHKEEPER will be fully integrated into the wider command and control digitised network passing data quickly to those that need it. The capability is being developed by Thales UK and will begin to deliver capability from the end of the decade.

Watchkeeper.

B8.37 Project **EAGLE** will provide an upgrade to the capability currently provided by the Sentry E-3D aircraft to enable it to carry out additional duties commensurate with an AWACS, such as control of aircraft and management of the air battle. The AWACS capability is critical in the successful prosecution of the air battle. This upgrade is required to bring the UK capability into line with NATO and US E-3 variants. The project is currently in the Assessment Phase.

B8.38 Project **HELIX** will provide incremental upgrades to the capability provided by the Nimrod R1 electronic reconnaissance aircraft and associated ground stations and training facilities. The programme is currently in the Assessment Phase.

B8.39 **UK INTELWEB** aims to determine the most cost effective solution for integration of intelligence systems to enable effective sharing and exploitation of intelligence material within MOD and with other Government departments.

B8.40 The Maritime Airborne Surveillance and Control (**MASC**) programme, aims to provide an assured airborne surveillance and control capability, forms the third element of the Carrier Strike capability, alongside the Joint Combat Aircraft and Future Carrier. It is envisaged to continue the capability currently afforded by the Sea King Mk 7 Airborne Surveillance and Control variant. MASC should provide a capability for surveillance of the air and surface as well as battle management of air defence fighters and other assets. The MASC programme is currently in the Assessment Phase.

B8.41 Established after publication of the Strategic Defence Review New Chapter, the **DABINETT** programme seeks to fill the challenging ISTAR capability gaps resulting from the changing threat. Focused on persistent ISTAR collection in the deep battlespace, the capability envisaged will be used to gather and disseminate strategic, operational and tactical intelligence, answer commanders' requests for information, and provide targeting information to systems in all environments. DABINETT will comprise a system of systems intended to enable persistent collection, processing and dissemination of near real time ISTAR data in the deep battlespace. A contract has been placed with LogicaCMG to undertake the role of DABINETT Development and Support Contractor (DDASC). This will support the Concept phase of the programme, establishing requirements and reaching a view on potential solutions, which are likely to roll out incrementally over the next two decades.

Illustrative planning assumptions

B8.42 We expect to spend in the region of £2BN to £2.5BN per annum in this sector over the next ten years. Figure B8(i) shows the Short Term Plan (STP) resource associated with the support of current in-service equipments and those that will enter service over the period. It also includes the resource associated with some service provision projects that are entirely funded from the STP. The figure does not differentiate between these two STP components.

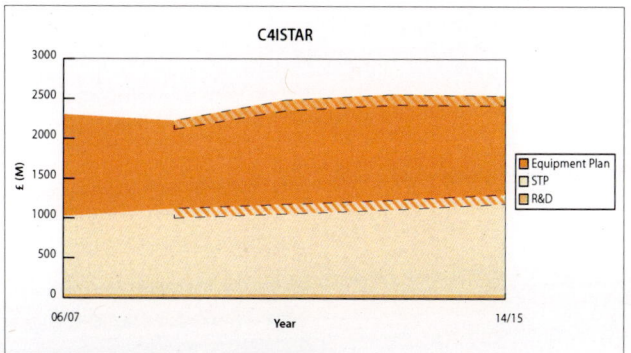

Figure B8(i) - Illustrative spend profile.
The above graph shows indicative spending in this sector over the next ten years. The figures from 08/09 are illustrative and include a range in order to emphasise the potential for shifts in investment priorities after the end of the current Spending Review period. This is prudent planning which does not distort the overall illustrative picture of general trends.

What is required for retention in the UK industrial base?

B8.43 In support of the forward programme of capabilities, the ability to design and manufacture equipment does not generally need to reside in the UK. However, there is a need to develop a cadre of system engineering skills to enable industry to understand our systems and in particular be able to support them through-life.

B8.44 In terms of specific capabilities for which a measure of sovereign control is required, information assurance and cryptographic systems protecting national eyes only information are particularly sensitive areas requiring a viable UK research, development, design, manufacture and support capability. A sustainment strategy must be developed in concert with other Government departments and is linked to current cross-Government information assurance (including cryptographic) studies.

B8.45   In other capability areas, the sovereign ability to design, demonstrate and perhaps build advanced systems may be advantageous to foster a balanced relationship with other nations. National capability in cryptographic systems, network integration and the design, build and operation of satellites, particularly small satellites, are examples of world-leading niche capabilities that may help to provide such leverage.

B8.46   More generally, across the C4ISTAR capability the UK needs to retain sufficient expertise to:

- understand capability requirements;
- develop user and system requirements;
- architect and maintain complex networks and communications with the ability to interoperate with a wide range of potential partners in support of operations;
- architect an overall system;
- assess the global marketplace;
- conduct research into areas that cannot be provided cost-effectively by the global marketplace;
- pull through successful research into in-service systems;
- test and evaluate systems;
- support within the UK equipment through life – including modification and technology insertion to facilitate upgrades and obsolescence management;
- maintain custody and the integrity of the increasing quantity of national military and intelligence data.

B8.47   In terms of technologies there are a number in which C4ISTAR are dependent, and in which there may, case by case, be a need for targeted investment to ensure a continued understanding of emerging developments or to have assurance regarding their security of supply. These include:

Antennas	Multilevel Security
Architectures	Nuclear C2
Combat ID	Radiation Hardening
Data Fusion	Radio Frequency (RF) integration and Electro Magnetic Capability
Digital Networks	Satellite Technology
Electro-optic/Infra-Red Imaging	Synthetic Aperture Radar
Electronic Counter Measures and Electronic Counter Counter Measures	Tactical Data Links
Electronic Support Measures	Sortware Defined Radios
Extra High Frequency satellite communications	Digital Signal Processing
High Integrity and Safety Critical Computing	
Wideband High Power Amplifiers	

Overview of the current UK defence market

B8.48   The C4ISTAR industrial sector is in good health. There is the potential for effective competition between onshore and offshore suppliers. Many of the companies within the sector have a strong presence in other markets, defence-related and civil, the latter providing opportunities for technologies to spin in to defence or to be exploited or further developed in support of exclusively defence demands. At the prime contract level, we recognise the following companies as particularly visible players at both Group and Division levels:

BAE Systems	**Thales**
EADS	**General Dynamics**
Lockheed Martin	**Northrop Grumman**
Raytheon	**Selex Communications**
VT Communications	**Ultra Electronics**
BT	**EDS**
Fujitsu	**LogicaCMG**
QinetiQ	

B8.49   At the sub-system level, the supply base is spread across a number of smaller enterprises as well as within the large primes, and often reaching into the commercial sector within the UK and overseas. Sourcing strategies reflect the general pattern of global supply and value chains exhibited across all high technology sectors. There is an increasing opportunity to develop innovative capability solutions across the supply chain from within traditional sub-system providers as well as those who may not necessarily consider themselves potential defence suppliers, for example mobile-phone data package providers or the biometrics industry.

Artist's impression of Skynet5.

B8.50   US investment and US companies dominate the world defence market and as the above list demonstrates, several of those US defence companies have made a substantial investment in a UK presence, attracted by the business opportunities and openness of the UK market. However, the larger US defence investment may create an imbalance in favour of US technologies in some important fields presenting a further challenge for sustainment, in particular regarding difficulties about technology transfer. Areas where the UK market presence has diminished in relative terms include EW and tactical communications.

B8.51   The wider civil market may offer a counter-balance for the future and provides significant potential for further innovation and competition. For example, the European mobile communications industry has highly developed digital networks and is a world leader. Civil IS providers have significant experience in the creation of secure and reliable networks which could have potential application in the military environment. Additionally, the ability of military personnel to rapidly absorb and act upon a wealth of information demands particular cognitive capabilities that could be provided by a variety of civil industry, which we may not have previously considered in depth such as film studios and computer console manufacturers who have significant experience and success in the wider cognitive field.

B8.52   The consensus of analysis of trends in defence expenditure is that C4ISTAR is a growing market. Factors driving this growth include:

- the massive market in electronic systems and software for domestic, commercial and entertainment purposes has created both the expectation and industrial foundation for a similarly rich military environment;

- development of new technologies, for example synthetic aperture radar, coupled with the contraction in sensor size that has enabled the expansion of those platforms able to carry ISTAR capabilities (e.g. on medium and large UAVs);

- application of commercial technologies. This is particularly true in the communications and 'personal devices' sectors, which have provided the technical basis for small radios, Global Positioning System (GPS) etc. Business tools, such as office automation, are also being used to support operations in a wide range of systems;

- there is rapid obsolescence which presents particular challenges and business opportunities for sub-systems that must be integrated into platforms with much longer life cycles. At its most extreme, this discontinuity of cycle can be seen in new platforms entering service with software and processors that are already at the margins of obsolescence. Therefore, management of spares and stocks will be critical to ensure these IS capabilities can continue to be serviced and updated throughout a potentially long service-life;

- doctrinal developments, such as the change in military structures from static to rapidly deployable and highly mobile forces that are flexibly 'networked';

- assimilation and dissemination of time-critical information to identify enemy forces and strike before concealment or withdrawal. This drives increased demand for sensor platforms capable of long-persistence and with the ability to relay information back to C2 systems;

- demand for greater interoperability with a 'coalition of the willing', which could include partners we have yet to undertake operations with. Lessons learned from Kosovo, Afghanistan and Iraq have led to a re-evaluation of requirements based on the need for interoperability with allied forces. Adoption of common standards within allied communication networks would facilitate greater interoperability, particular coupled with the use of Commercial-off-the-Shelf (COTS) products, where appropriate, amongst likely coalition partners;

- budgetary pressure to leverage additional capability from existing platforms and to upgrade existing C4ISTAR equipment;

- traditional boundaries between land, sea and air forces are blurring – network-enabled forces will further reduce the divide and also means that systems must interoperate across all environments, fusing outputs from multiple sensors.

B8.53   Although the vast majority of equipment expenditure is currently via the established defence contractors, much of the systems content is based upon exploitation of technologies derived from commercial communications, computing, signal and data processing capabilities. While many skills are applied to specific defence needs and constraints, the individuals in possession of those skills are readily deployable outside defence. Defence programmes also provide interesting and innovative development work, although expertise can quickly waste and transfer to commercial applications if there is insufficient business. This is a critical consideration for sustainment, as work to modify long-serving systems is likely to be constrained by intellectual rather than technical resource. This is more significant where there are challenging defence characteristics, for example packaging for harsh environments, extreme security protection, and extreme reliability. Indeed, the boundaries between the 'business' and the 'battlespace' environments are blurring, with what might be considered as 'office' applications and services being delivered close to the front line.

Sustainment strategy

B8.54   We require to have available to defence a comprehensive understanding of rapidly developing technologies and the ability to exploit these in conjunction with existing systems, with which they must interoperate. We have identified four areas of industrial capability within the C4ISTAR sector that are essential for the maintenance of UK operational sovereignty. These are:

- high grade cryptography and associated information assurance capabilities;

- a continued ability to understand, integrate, assure and modify mission critical systems;

- a continued ability (for both us and industry) to act as an intelligent customer, in particular to track emerging technologies for potential military application;

- a sustained research and development base (attracting the right calibre of individuals with the right skills) supported by a manufacturing capability in specific areas of defence technology.

B8.55   Of these, only maintaining a cryptographic capability currently requires a strategy to sustain an end-to-end design, development and manufacturing capability. This derives from a need for the design, development and production teams to be comprised of UK Nationals, and in part because there are limited commercial opportunities for this technology. The maintenance of other C4ISTAR capabilities resides currently within a relatively strong set of industrial primes supported by a diverse supply chain. Of these, companies with a commercial, as opposed to those with a traditional defence sector background are increasingly offering themselves as players in the system engineering arena.

B8.56   But we must also change the way we present our requirements to, and interact with, the market. The relative strength of the C4ISTAR industrial base and the complementary technology acceleration within the commercial sector are opportunities which we intend to exploit in support of our drive towards NEC.

High grade cryptography

B8.57   Although the market for high grade cryptography is low in volume compared with the commercial sector's lower grade cryptography needs, it remains strategically vital across Government. Protection of high grade information for MOD and wider Government requires a UK sovereign capability to control those aspects of cryptographic production and support critical to the integrity of the product and protection of issues such as national security. The broader benefits of inclusion and access to developments in the wider cryptographic community (in this case predominantly the US) will also be pursued. In terms of quantity, we represent the major customer for high grade encryption within Government, and as such have a significant role to play in shaping the grade and type of cryptography offered by industry in the future. However, our interaction with the market is not aided by the fact that the Government's demand for cryptography is currently fragmented, thus depriving industry of an aggregated basis which to develop a sustainable business plan.

B8.58   Work, supported by us, is being led by the Communications Electronics Security Group (CESG), part of the Government Communications Headquarters (GCHQ), to assess this aggregated demand and generate better coherence across Government. In essence, the objective of the work will be to generate a "pyramid of demand" that could put the demand for UK design and manufacture of high grade cryptography alongside a consolidated pan-Government demand for a lower grade product.

B8

C4ISTAR

A Private from 2nd Battalion The Parachute Regiment with a Personal Role Radio on security patrol in the centre of Kabul Afghanistan.

in conjunction with industry via programmes such as NITEworks[2] and DABINETT and the establishment of the Integration Authority but the approach needs to be further developed into a coherent programme delivery doctrine that includes all phases from concept to disposal.

Research and technology

B8.62   In order to sustain a vibrant onshore R&T base in this sector, both within MOD and industry, we are maintaining an overall research programme in this area currently funded at around £55M per annum. Figure B8(ii) illustrates the proportion of the programme currently allocated to each of the major elements of interest. This allocation will change as research priorities are adjusted to respond to developing areas of interest, but it does reflect a key aspect of our ability to remain as an intelligent customer of emerging technologies and concepts.

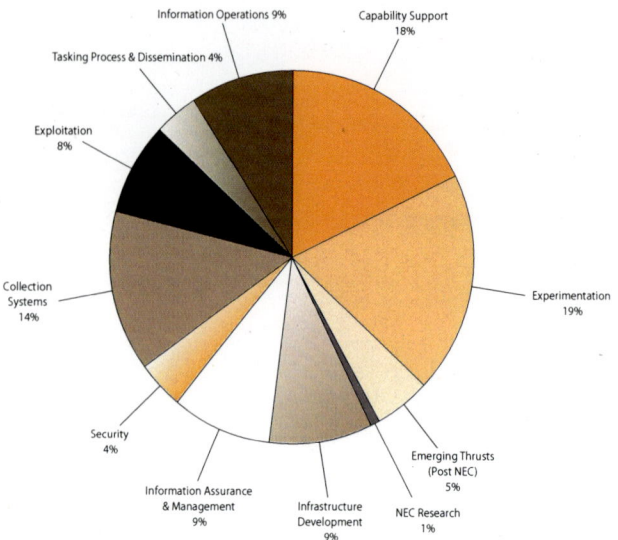

Figure B8(ii).

System engineering and assurance and the 'intelligent customer'

B8.59   Availability of, and assured access to a systems engineering skill base is critical to deliver the vision of NEC. We will need to maintain an intelligent customer status, which will not be easy, given the pace of change in the industrial and technology base dictated by the commercial market.

B8.60   Unlike the cryptography market there is a good spread of competent systems engineering organisations within the UK and the global industrial base. But set against that, the overall health of the C4ISTAR defence sector is complicated by the fact that many of the underpinning technologies for C4I in particular are also relevant to a much larger global commercial market. Therefore we are in general a relatively minor customer in a market where the pace of technological change creates its own set of unique pressures for capability sustainment. The risk for us is not generally that the capability is lost due to lack of demand but that the pace of change in the commercial world, move a company's capabilities away from the technological direction we require. To mitigate this, we must ensure that the profile of our programmes, and the potential earnings yield is such as to allow companies to use defence as well as commercial drivers to innovate and climb the technological staircase. In this way, industry will be motivated to sustain sufficient skills sets to maintain and support our systems. This will require greater attention to the profit-risk mechanisms that we apply to our contracts.

B8.61   Finally, the ability to retain an understanding of the full system in order to provide through-life support must be considered. In a number of instances, it would not be appropriate for C4ISTAR equipments to be supported offshore, especially where these include software modifications that might reduce confidence in information assurance. We have started to create and develop the understanding of the full NEC system

B8.63   It is neither feasible nor affordable to sustain a critical capability in all of the technologies of potential interest. However, we can expect that much of the underlying technology and applied expertise will be driven by the commercial marketplace and international defence investment.

B8.64   This will demand closer interaction with industry. Applicable technology tends to be exploited across a number of projects, and companies are constantly seeking to add technical value to improve their competitive advantage and product differentiation. Private venture investment could be augmented by our funding to advance knowledge and engineering expertise. We would need some guaranteed return (as would the company), if the chosen technology proved to be applicable to the target project. We may also wish to make resulting IPR more accessible.

B8.65   We intend to examine whether this approach might be realised via an extension of the Defence Technology Centre concept. The associated MOD-industry consortium would need to maintain a view of future technology, and of future programme requirements, in order that future projects could benefit from timely investment in new and improved components. The selection of technologies and development of applications would inform applied research and build a cadre of expertise, reducing risks and timescales, and enhancing UK competitiveness.

---

[2] *a unique experimental environment that allows us to assess the benefits of NEC, the optimum use of military systems and options for effective and timely delivery of NEC*

Presenting our requirements to the market

B8.66   We will provide increased transparency of our capability intentions (within obvious security considerations), clarity on affordability constraints and better articulation of the desired enterprise outcome.

B8.67   We will also ensure coherence in our approach to the overall NEC architecture, where this is relevant, and will share this with industry - from high-level overviews, to more detailed expressions at functional or capability level. The Ministry of Defence Architectural Framework (MODAF) will be key to expressing NEC requirements in consistent terms and contributing to the development of genuinely joined-up doctrine. This will aid comprehension of intent between us as customer and industry suppliers, and highlight key areas to be addressed in the development of solutions such as interfaces, data specifications and protocols.

Open Systems Benefits

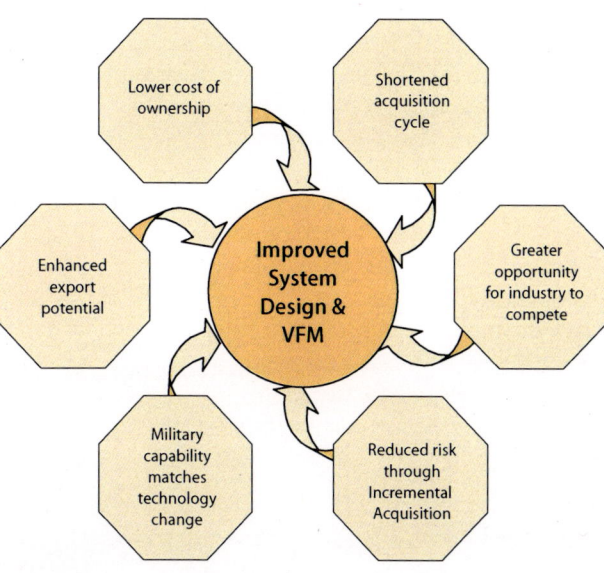

Figure B8(iii).

MODAF

It has been recognised for some time within the MOD that a consistent means of defining platform and system interfaces across all MOD acquisition activities is a key enabler to achieving the interoperability that is at the heart of Network Enabled Capability. Therefore, a methodology for developing consistent architectures across the MOD was developed, this methodology is MODAF. The primary objectives of the implementation project are:

- To develop an Architectural Framework based upon defence and commercial best practice and to tailor this to MOD-specific acquisition and operational processes.
- To develop the directives and guidance necessary for this architectural framework to be implemented.
- To make recommendations to vendors such that suitable tool sets are developed and made available to all of the parties that may utilise the architectural framework.
- To embed the business change required to implement the architectural framework such that it becomes self-sustaining without continued input from the implementation project.

B8.68   Complementing this, we will wish to see solutions that make the greatest use of open systems architecture approaches. The use of interface standards that are non-proprietary, already established and recognised by formal bodies, or which are accepted as de facto standards by the market, should enable industry to develop solutions that maximise commercial opportunities and assist future export potential (see figure B8(iii). The challenge for us will be to avoid a shift to our simply being a consumer with insufficient influence over system design. For some requirements we may still need closed or only partially-open systems.

Conclusion

B8.69   C4ISTAR capabilities are essential to the achievement of NEC. In support of this endeavour, the C4ISTAR industrial sector is broadly in good health with significant potential for enduring earnings across defence and commercial customer bases. The forward C4ISTAR equipment and associated support programmes represent a sizeable proportion of our forward plans, recognising the priority we attach to delivering NEC. Beyond certain high classification requirements, our interests lie principally in the sustainment of a highly skilled sector capable of undertaking complex systems integration activities and supporting information assurance initiatives. This is consistent with broader Government objectives to support the further development of high technology industry and research in the UK.

B8.70   We can in general benefit from the extended customer base and the weight of civil investment it attracts. But there will also be a continuing need to maintain awareness of the depth and breadth within the UK industrial base of those skills necessary to meet and support high-end defence requirements. In those areas there is a risk that unless the skills are exercised regularly, they and/or their currency will diminish.

B8.71   So the strategy for C4ISTAR will be to:

- achieve better cross-Government coordination of demand for cryptography;

- work with all areas of industry to target defence and commercial research expenditure to activities that offer the greatest potential Defence benefit and which have clear exploitation paths;

- continue to encourage a wider civil industry to explore the potential application of its knowledge and products to the defence market;

- give industry visibility of our forward plans, and where appropriate the opportunity to help develop potential solutions from an early stage;

- identify intended system architectures and associated interface details;

- encourage and support the use of open systems architectures and COTS products to the greatest practicable degree.

Definition

B9.1   The Chemical, Biological, Radiological and Nuclear (CBRN) protection capability area is focused on the development and provision of capabilities that support our CBRN defence policy. These capabilities are broken down into three broad categories:

- **Timely warning** - the detection and identification of CBRN weapons and materials, with the ability to process data and provide information to facilitate decision-taking and action;

- **Survive** - the capabilities, based around the person, necessary to survive a CBRN challenge: and

- **Sustain** - the unit/force level capabilities required to, confirm the extent of a hazard, rapidly recover from an event, and sustain/regain operational tempo.

Strategic overview

B9.2   We are committed to maintaining the UK's political and military freedom of action despite the presence, threat or use of CBRN weapons. The spectrum of this threat is broad and may come from a number of different aggressors, from state-sponsored programmes to international terrorist organisations. The 2003 Defence White Paper, 'Delivering Security in a Changing World', identified international terrorism and the proliferation of Weapons of Mass Destruction (WMD) as 'the two most serious and direct threats to our peace and security'.

B9.3   As the CBRN threat becomes more diverse, the defensive CBRN capability required by today's service personnel inevitably broadens. Our Armed Forces are able to deploy rapidly to almost any region of the world, and face an increasing number of potential CBRN challenges. They may come across remnants of discontinued CBRN programmes, encounter endemic diseases that could be difficult to distinguish from biological warfare (BW) attacks, or be exposed to toxic industrial hazards (TIHs). In all cases, having the right capability is absolutely key to success.

B9.4   With the need for flexible rapidly-deployable Armed Forces that are able to operate in a number of different circumstances, our force protection posture requires similar flexibility and should impart a low physiological burden within a minimal logistical footprint. We are therefore developing and enhancing integrated surveillance, warning and reporting, monitoring, protection and detection capabilities. Fundamental to the future of CBRN protection are the opportunities represented by Network Enabled Capability (NEC). Intelligent networking will give increased situational awareness and facilitate rapid hazard prediction. Data fusion from a multitude of sensors will create the CBRN recognised picture.

B9.5   Aligned to the rapid technological development in this sector of defence are the potential economies of scale to be found when considering the wider CBRN market place. In particular this includes

Pre deployment CBRN and damage control training on board HMS INVINCIBLE.

the homeland defence/counter terrorism sector which is led by the Home Office and the consequence management area through the Department of Health. A major strand of our industrial strategy is to fully investigate how these economies can be exploited and we are fully engaged with both the Home Office[1] and the Department of Health[2].

B9.6   Strong CBRN force protection has always formed part of the UK's deterrent posture. We must equip our highly professional troops with advanced equipment able to readily detect, withstand and respond to CBRN challenges. Only by achieving this can we send the powerful message to all aggressors that the impact of any CBRN attack will be low and never achieve its goal.

Equipment programmes

B9.7   The UK's CBRN protection programme is focused on rapidly developing emerging technology to deliver solutions to the complex threat environment. The threat we face is almost limitless, from the more conventional industrially manufactured weapons systems to the emergence of global terrorism. To combat this significant technical challenge there are two main themes to the programme. The first is a strong research base that seeks to characterise the threat and establish the principles to counter that threat. Second is an incremental acquisition approach to deliver effective systems rapidly to the front line. Our targeted research, therefore, drives industrial development, supports growth of intelligent suppliers and influences priorities for investment. For this reason our research establishment has been retained within Dstl, as a strategic asset.

B9.8   We take a systems approach to CBRN protection. The ability of industry to deliver fully integrated systems, often from a wide variety of technology suppliers is key. At the heart of the programme is NEC which will enable rapid decision making and increased situational awareness. The CBRN programme is divided into the conceptual groupings of Timely Warning, Survive and Sustain.

B9.9   Outlined below is the current CBRN equipment programme. Over the coming months the programme is undergoing a review to enable more effective capability management. It is the precursor to facilitate timely incremental acquisition and the rapid exploitation of emerging technologies. It will entail the brigading of the current programme assumptions under **Timely Warning, Survive** and **Sustain** and will probably recast the manner in which these assumptions are delivered. The net result is that the assumptions remain valid, however, the project name and delivery profile may change.

B9.10   **Timely Warning** – the capabilities required to 'warn the force… in time' in order to avoid, to take decisive action, to protect:

- Warning and reporting is a fundamental part of CBRN planning and deterrence and is dependent on other command and control systems, an agent database and national threat assessments. These elements are, of necessity, classified and their details must remain within the UK. The **NBC Battlefield Information System Application (BISA)** is the software providing CBRN input to the networked battlespace, helping achieve NEC. It will be integrated with other defence applications and enable operational decisions to be made in anticipation of, or response to, a CBRN event. Increment 1 is planned to be delivered in 2007. A further increment with increased functionality is planned and should follow within five years. It is expected to have a service life into the 2020s.

[1] Engagement is through the Home Office chaired CBRN Strategic Board, Delivery Board and Research Strategy Board.
[2] Collaboration with the Department of Health and its purchasing organisation to enhance national capabilities. In addition efficiencies are gained through stockpile sharing and joint contracting.

- The **Integrated Sensor Management System (ISMS)**, which will provide a 'plug and play' suite of sensors to Land and Air forces, will be one of the systems providing the BISA with information to enable timely warning of a CBRN event. The system, which will be able to interface with in-service chemical, radiological and nuclear detectors, is to be delivered with a commercial off the shelf (COTS) generic unmanned biological detection system and is planned to enter service in 2006 and to have a service life into the 2020s.

- UK military and civil biological detection and identification capability is a fundamental part of basic force protection. The **Unmanned Biological Detection System (UBDS)**, which will be available by the end of the decade, and will provide generic bio-detection to complement the ISMS programme. It will function in concert with other in-service detectors and provide a base upon which **Biological Detection Tier 3 (BDT 3)** can develop.

- **BDT 3** will combine stand-off detectors and simple sensors to provide a detect to warn capability. With future integration opportunities a possibility, BDT 3 may replace some planned Detection, Identification, Monitoring and Analysis (DIMA) systems, such as UBDS. The system is expected to be available by the middle of the next decade.

Man-portable Chemical Agent Detector.

- With a planned fielding date for the end of this decade the **Maritime Biological Detection System (MBDS)**, will provide the Royal Navy (RN) with a fully automated biological detection and identification (D&I) capability that will give an increase in capability with a reduction in the logistical burden associated with current D&I systems.

- The **Lightweight Chemical Agent Detector (LCAD)** and **Man-portable Chemical Agent Detectors (MCAD)** in-service since 2004 provide alarm at attack levels of Chemical Warfare Agent (CWA). Both utilise **Ion-Mobility Spectrometry (IMS)** which is a pivotal technology to current capability and provides the foundation for the growth of future systems and are expected to have a service life of at least 15 years.

- **Stand Off Chemical Detector (SOCD)** has a proposed delivery date in the middle of the next decade and will provide a stand off liquid and vapour detection capability. It provides timely warning (an alarm) and sustain (reconnoitre and survey) capabilities. It is envisaged this will be a key component of future defence platforms.

- The Joint CBRN Regiment's **Light Role Teams (LRTs)** will provide a strategic CBRN detection, identification, monitoring, analysis and transportation capability matched to early entry light forces for rapid deployment. These specialist 8-man teams also provide the operational commander with expert CBRN advice

B9

CBRN Force Protection

and ability to prepare the way for follow-on forces. Expected to be available in the latter half of this decade, the teams will be self-sufficient and have reach-back capability to the UK.

● Dstl sponsored research on reagents is directly feeding into the **Rapid Diagnosis System (RDS)** which is being developed to identify illness that may have been caused by undetected low level exposure to possible BW agents, TIHs or endemic disease. Anomalous symptom events will be detected by the **Real-time Medical Surveillance System (RMS)** using software analysis to assist the man in the loop.

B9.11 **Survive** – the capabilities, based around man, 'to survive a CBRN challenge':

● **Personal NBC Protection System (PNPS)** will, by the middle of the next decade, provide all personnel with an integrated system of Individual Protective Equipment (IPE) that gives protection against a range of NBC hazards with a reduced physiological burden. It will be aligned to the future soldier systems (such as Future Integrated Soldier Technology, FIST) which may affect the planned ISD. It will be used in conjunction with the **General Service Respirator (GSR)** which is due in-service by the end of next year. The GSR will provide very high levels of respiratory protection while reducing the physiological and psychological burden suffered by personnel using this equipment and is expected to have a service life well into the 2030s.

● From the middle of the next decade, the **Deployable Collective Protection System (DCPS)** will provide an easily transportable and erectable collective protection system. It is an aspiration to combine this with standard military tentage, where considerations include coatings, filtration and decontamination.

● The incrementally acquired **Aircraft and Aircrew CBRN Survive to Operate (AACSTO)** programme looks to provide protection and hazard management capability in aircraft operations. It seeks to reduce the burden on the individual, improve decontamination of aircraft and protect the insides of certain aircraft from cross contamination. Due to be phased in service from the middle of the next decade, the programme represents a significant enhancement to our current capability.

B9.12 **Sustain** – the unit/force level capabilities to confirm the extent of a hazard, to rapidly recover from an event, and to sustain/regain operational tempo:

● The **NBC Reconnaissance & Survey (NBC R&S)** programme is designed to develop the capability gap left as Fuchs (reconnaissance vehicle) approaches the end of its life. Planned for introduction in the middle of the next decade and currently planned as a FRES (Future Rapid Effects System) variant which will deliver a full suite of CBRN detectors.

● Using a multitude of systems to monitor a wide variety of surfaces, including personnel, equipment, buildings and terrain the **Chemical Monitoring Capability (CMC)** is currently split into two programmes. The **Surface Detection System (SDS)** which aims to confirm and identify the presence of known threat chemical agents will be used in conjunction with an **Unmasking Aid (UA)**. The preferred solution may involve the use of disclosure technology, IMS and flame photometry. **Biological Surface Monitoring Capability (BSMC)** will provide a biological equivalent to SDS. Both systems are planned to enter service in the middle of the next decade.

Troops based in Qatar, practice masking-up drills during an NBC drill.

● **Multi Level Decontamination (MLD)** is aimed at providing a 'thorough' decontamination capability at unit level through the joint operational area, from personnel to sensitive equipment and whole platform and is expected to be available in the middle of the next decade.

● **Unit Decontamination Capability (UDC)** is a programme to provide a low burden integrated self-contained 'thorough' decontamination capability to enable the removal of IPE and reduce casualties. Planned to be available by the latter half of this decade, it may become part of a system of systems within the longer term MLD programme.

● **Aircrew Chemical Agent Detector (ACAD)** will provide a chemical detection capability with unparalleled levels of sensitivity from the middle of the next decade. It is an aspiration to merge ACAD with the replacement unmasking aid to deliver a widely deployable capability which upgrades/replaces LCAD.

● The recently delivered **Tactical Radiation Monitoring Equipment (TRaME)** provides a comprehensive suite of detectors and monitors for our use on the battlefield. In addition sensors, for a number of current and future platforms, are being researched in conjunction with the host platform's development. Therefore, those platform based detectors fall outside the scope of this chapter. TRaME is in-service now and is expected to have a service life of into the 2020s.

● **Medical Countermeasures (Med CM)**. Whilst protection from, and avoidance of Biological and Chemical attack may be possible, there are many scenarios in which our forces could be exposed to biological and chemical agents. In these instances effective Med CM will be essential in avoiding casualties and maintaining operational tempo. Med CM can be delivered either as pre or post event prophylaxis, or as post exposure therapy.

● The MOD has a vigorous research and development programme in this area based around the widely recognised excellence and expertise of Dstl and is addressing nerve agent antidotes, vaccines and antitoxins. Within the next five years we plan that an improved nerve agent antidote, a licensed antitoxin and a number of vaccines will have been fielded. In the subsequent decade it is anticipated that further countermeasures will be made available.

Indicative planning assumptions

B9.13   Figure B9(i) illustrates our current assumptions for CBRN spend over the next 10 years. This includes the planned spend on the Equipment Plan (EP), the cost of maintaining and operating the equipment captured within the Short Term Plan (STP) and the planned research funding to identify and develop new technologies to support the EP. The UK will spend around £120M on CBRN protection acquisition, support and research next year. The total CBRN package is planned to increase out to 2015 as a number of the large programmes are delivered. We will continue to focus on developing new technical solutions, whilst maintaining and improving our existing capability. The CBRN programme described above does not include the investment made as part of the platform based programmes.

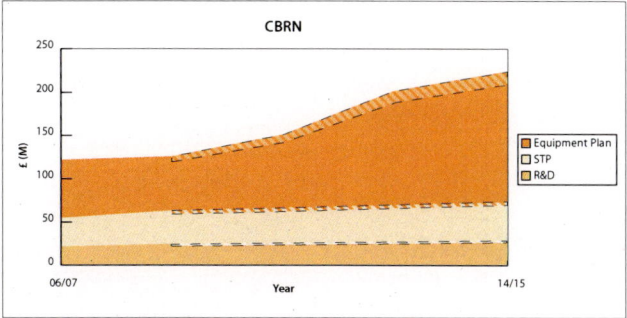

Figure B9(i) - Illustrative spend profile.
The above graph shows indicative spending in this sector over the next ten years. The figures from 08/09 are illustrative and include a range in order to emphasise the potential for shifts in investment priorities after the end of the current Spending Review period. This is prudent planning which does not distort the overall illustrative picture of general trends.

What is required for retention in the UK industrial base?

B9.14   The CBRN capability area and the associated industrial sector are distinctive in that the UK's posture is purely defensive. In addition the UK is a State Party to a number of non-proliferation treaties[3]. This governs the breadth of the market place for our industrial partners often limiting the scope of R&D activities, what can be sold, and also to which countries. The global commercial free flow of sensitive technology is further limited by our national security concerns and those of our international allies. Therefore, the need to retain a technological advantage in the field aligned to our national strategic priorities drives us to retain a strong on-shore focus.

B9.15   The UK is at the forefront of the development and production of CBRN protection systems. This is in the main due to an exceptionally strong research programme in conjunction with the UK's successful development and manufacturing base. This places the UK in a unique position with its strategic partners. In particular, the UK can achieve very advantageous research and technology partnering through international agreements and as such exerts significant strategic influence.

B9.16   Providing effective CBRN protection is a constant battle to develop technological solutions to complex problems. Taking advantage of rapidly developing technology, incrementally, to continuously increase our capability is essential, in conjunction with a focused and agile industrial sector. Maintaining this capability on-shore will result in a supplier base which fully understands the UK's approach to CBRN and is not distracted by other nations' strategic goals. It is therefore vital to guarantee access to the following key functions in the CBRN area in order to enable the delivery of military effects to the Armed Forces:

- Exploit emerging technology and developing equipment/systems;
- Control CBRN technology proliferation[4];
- Maintain a national CBRN focus to influence across defence[5];
- Exploit economies of scale across Government;
- Respond to Urgent Operational Requirements (UORs);
- Retain credible involvement in CBRN deterrence;
- Maintain security of supply.

B9.17   In order to deliver these key functions we need access to an industrial base that, as a minimum, is capable of providing a world-class capability in the following broad areas:

- **Intelligent Supplier**. Due to the rapid pull through of technology and the complexity of the CBRN area, our prime concern is the retention onshore of the intelligent supplier who should have in-depth knowledge of CBRN in terms of the technical solutions available and the context in which we will use them. This concept underpins the relationship we have with industry, guaranteeing supply and ensuring both our Armed Forces and industrial base remain at the forefront of technology. The intelligent supplier should be comfortable working closely with the defence research base to develop principles and deliver solutions.

- **System Engineering**. CBRN systems are complex and very few contractors possess all the skills to deliver the complete solution purely from in-house resources. Vital to us is the retention onshore of the ability to Systems Engineer. The component parts may well be sourced offshore, however, the ability to deliver the total package, inclusive of those security sensitive items is key. This will allow the development of systems that suit our military doctrine and whose procurement strategy is not affected by other nations' strategic goals.

- **Technology Exploitation**. The sensitivity of specifics of the CBRN threat and a lack of demand in the commercial sector drive us to invest a significant amount in R&T; the output being a proof of principle or the identification of a technology that with further development would deliver a significant increase to the capability of CBRN protection. This development is only possible using the Intelligent Suppliers mentioned above. Technologies of particular note that must be retained and developed on-shore are those associated with: **Chemical Detection and Identification, Biological Agent Detection and Identification, Physical Protection** and **Warning and Reporting**. In all these areas the UK specific protection factors and the underpinning intelligence to support them makes the use of offshore suppliers undesirable. In addition the strategic priority associated with CBRN makes the reliance on a third country for the development of technology impossible. The UK is a world leader ion-mobility spectrometry and the maintenance of this capability in the UK gives a special influence in collaboration. The arguments for not allowing defence capabilities to be dependent on the availability of technology and manufactured products from overseas are equally valid in this area.

---

[3] *Chemical Weapons Convention and Biological and Toxin Weapons Convention.*

[4] *Non-proliferation and export licences are considered fundamental to continued development of the industrial base while protecting key technologies and the military advantage - this provides a win-win for us and industry. This extends from the development of Dstl research, through provision of a protected supply chain to disposal.*
[5] *CBRN technologies and thinking must be integrated with overall military capability. It is intrinsic to all areas and, as a common enabler, the requirements for CBRN protection must be considered at the earliest stages of all acquisitions.*

- **Supply of Raw Materials**. In conjunction with the need to ensure security of supply of technology is the need to ensure an onshore manufacturer that can provide certain key raw materials. It is our intent to reduce the reliance of large stockpiles and engage in procurement based on surge production. Of particular note in this respect is the supply of **Anti Gas Cloth** for CBRN protective suits where we must be able to guarantee a product that meets our stringent quality and protection requirements. Certain other raw materials are fundamental to the UK's **biological defence** capability and their security, quality and guaranteed supply must not be compromised. The ability to surge, with the associated logistic implications will be carefully considered on a case by case basis.

- **Medical Countermeasures Manufacture**. Medical Countermeasures are the most sensitive area of our CBRN interests. The capabilities of the countermeasures and the definition of the threat that they counter is of such a sensitive nature that sole reliance on offshore development and manufacturing is impossible. Certain agreements do exist with our close Allies to share research and in some instances collaboration in procurement exists, however, the retention of a strong UK skills and manufacture base is vital. This industrial process is also of significant interest to other Government departments.

Overview of the current global and UK market

B9.18   CBRN protection requirements have for some time been met through a healthy and competitive industrial market place. All of the current larger value contracts have been established through competitive tendering exercises with, notably, no single contractor more successful than the other main competitors.

B9.19   The supplier base within the UK is strong and growing due in part to increased focus on the homeland defence market. This is attracting suppliers who have not previously shown a defence interest, which should benefit us by facilitating access to innovative solutions and technologies. The industry is represented by NBC UK, a special interest group within the Defence Manufacturers' Association. The membership of 55 companies has a significant interest in CBRN either as integrators, manufacturers in their own right or as sub-contractors. Those represented span every field from CBRN protection for the individual (clothing) via man-portable detection equipment right through to large highly specialised systems and equipments for the detection and identification of biological agents. Currently UK industry has interests in the home market as well as the increasingly profitable international sector[6]. The industry remains firmly focused on meeting domestic requirements; however, although the UK market is buoyant and our predicted spend is likely to increase, the US market continues to present an attractive commercial proposition to industry. If industry fails to achieve its commercial expectations within the UK, its focus may shift from not only exploiting US opportunities as they arise, but diverting research and technology investment into that market. UK industrial consolidation and development into the US would therefore present a risk to us associated with technology transfer and could harm our operational effectiveness.

B9.20   Although the number of companies interested in CBRN business is large, often the contractors are relatively small and specialist. There are only a few companies that could be considered targets as industrial partners or capable of providing a prime contractor role. These are currently: **Smiths Detection** (detection and identification), **General Dynamics UK** (systems integration and manufacture), **SERCO Assurance** (systems integration), **EDS** (systems integration).

---

[6] Example: Smiths Detection has recently made inroads into the Japanese market. The industry's most profitable sector is the US marketplace, both military and homeland defence.

The Joint CBRN Regiment capabilities include nuclear and chemical survey, biological agent detection, and post attack decontamination .

Sustainment strategy

B9.21   The UK's commercial CBRN sector is buoyant. To manage effectively both MOD's and industry's aspirations the following strategy will be followed:

- Incrementally acquire to deliver solutions rapidly and avoid the feast or famine cycle for industry.
- Explore innovative partnering opportunities to protect and develop the priority capabilities and processes we wish to retain in the UK.
- Continue to focus research on the early transfer of technology to industry.
- Maximise the economies of scale from effective cross Government working.
- Early engagement with industry to establish how best they can achieve their global commercial aspirations within the boundaries imposed by the use of UK specific sensitive technologies.

B9.22   The CBRN procurement strategy is thus primarily one of competition that seeks the best candidate to ensure value for money whilst guaranteeing performance levels. Whole life cost is a particular concern as is the reduction of risk. Whilst the onshore industrial base is buoyant and generally able to deliver, offshore suppliers will not be precluded unless their inclusion increases risk, in particular those based around security, security of supply and protection of sensitive information. We recognise that to maintain an 'intelligent supplier' base in the UK some of the development, production and support contracts will have to be targeted at UK based industry. International Agreements on non-proliferation will also be honoured. We will explore however the potential costs and benefits of partnering, particularly with the four main industrial players in the UK, to see whether other acquisition models could allow us to achieve rapid and innovative acquisition alongside better value for money.

B9.23   To protect what we assess as our priority capabilities, should the marketplace become less buoyant, CBRN research within Dstl has been retained as a strategic asset. This enables the underlying science to be progressed to a point where technology transfer to industry can take place. Therefore, research across the spectrum of Dstl's activity is focused on the delivery of technical solutions to the CBRN problem, which is readily available for industrial exploitation. If necessary Dstl can produce fieldable solutions in niche areas where there is no commercial market.

The development of the world's first regeneratable NBC filtration systems now coming into production for the UK's TITAN and TROJAN engineering tank systems and the TERRIER battlefield engineering vehicle, were a direct spin-in of commercially funded research and development. These systems,were developed by Domnick Hunter Ltd for TITAN and TROJAN, and by the combined efforts of Ametek, Aircontrol Technologies and Pall for TERRIER. The systems were de-risked by the use of MOD funded Technology Demonstrator Programmes (TDPs), and we see TDPs as an important mechanism in the future.

Way ahead

B9.24   We will analyse the opportunities available within wider Government to derive economies of scale for equipment purchased across Departments, building on the existing dialogue between MOD, the Home Office and the Department of Health. This will include investigating the potential for co-operative purchasing, shared research, stockpile sharing and joint contracting. Additionally, whilst we have been successful in utilising competition as a procurement strategy in the past, we want to work with industry, in particular those companies who are members of NBC UK, to look at the benefits of other models, such as partnering, for acquisition to ensure continued value for money whilst maintaining an innovative and flexible supply-chain. This work will begin in early 2006.

Definition

B10.1   Counter terrorism (CT) is a pan-Government activity, the policy responsibility for which sits primarily with the Home Office (for activities internal to the UK) and the Foreign & Commonwealth Office (for activities overseas). The MOD provides military support, and takes the lead, in those areas where military forces on operations are at risk from terrorist attack.

Strategic overview

B10.2   The threat from international terrorism to the security and economic interests of the UK, its partners and its allies will, for the foreseeable future, derive from individuals and networks motivated by extremist ideologies which are committed to an international campaign against the West and those nations and governments associated with Western ideals or interests. Their activities will be marked by an extreme ruthlessness in seeking to maximise civilian casualties. Suicide attacks will be a feature of their operations. They will exploit as mass-effect weapons readily available means such as aircraft and tankers, and they will aim to develop Chemical, Biological, Radiological and Nuclear (CBRN) means of attack.

B10.3   The threat presented by terrorist networks will be enduring, although the level of support and sympathy for their objectives from host nations will depend upon: the degree of economic prosperity imbalance with the West; the impact of reform and progress towards democratisation; the effectiveness of CT operations; progress with key issues such as the Middle East Peace Process; and the scale of cultural antipathy towards the West.

B10.4   Given the nature of the threat, capabilities traditionally needed in niche areas and Northern Ireland are increasingly becoming required across the force structure. This reinforces the importance of the counter terrorism sector, a sector to which the UK is a major contributor, and provides greater opportunities for both industry and MOD to become more cost-effective in the CT field.

Equipment programmes

B10.5   The development and procurement of specialist CT capabilities is largely led by the Director of Equipment Capability (Special Projects) – DEC (SP) – and the Special Projects Integrated Projects Teams (IPTs) within the Defence Procurement Agency. Procurement projects are currently being run in the areas of mobility, communications, force protection and improvised explosive device detection and disposal, amongst others. The details of these projects remain classified, and are contained in a separate and expanded version of this chapter, the details of which are available to suitably qualified companies at the MOD's discretion.

B10.6   Many of the projects being run by DEC(SP) and the Special Projects IPTs have utility outside the CT arena, and could, for instance, support general Special Forces (SF) operations in wartime.

Indicative planning assumptions

B10.7   CT requirements are captured within the Capability Area Plan (CAP) for DEC(SP) within the MOD, and drive both the research into and procurement of new equipment.

B10.8   Figure B10(i) illustrates the currently assumed overall resources for the CT Equipment Programme, Short Term Plan support costs and CT research over the next decade.

B10.9   It is worth noting that the MOD's spending in this sector have been bolstered over recent years by spending on Urgent Operational Requirements. The UK will continue to require the wherewithal to rapidly advance new technologies into the front-line.

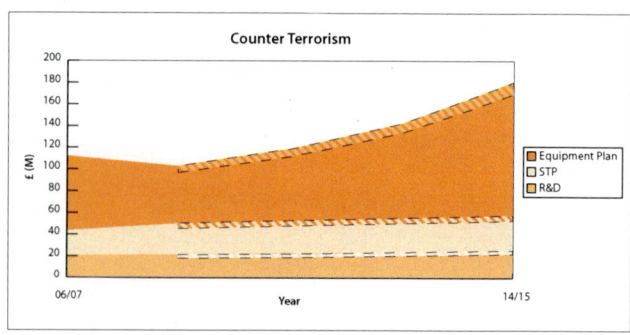

Figure B10(i) Illustrative spend profile.
The above graph shows indicative spending in this sector over the next ten years. The figures from 08/09 are illustrative and include a range in order to emphasise the potential for shifts in investment priorities after the end of the current Spending Review period. This is prudent planning which does not distort the overall illustrative picture of general trends.

Capability priorities

B10.10   In contrast to many of the areas covered by the rest of the DIS, CT is underpinned by technologies utilised in many areas of industry – from the cutting-edge physics of electro-optics and RF sources through to the rather cruder, though no less important, problems of vehicle mobility and specialist demolitions charges. In some cases details of the technology need to be protected, as their exposure would reveal critical military capabilities. In other cases the technology used is well understood, but the particular way it is implemented within a system, or the way it is used on an operation, remain sensitive.

B10.11   Although there are aspects of the technology base within the development, manufacture and sustainment of a CT system that need to be retained within UK industry, it is primarily within the areas of systems engineering (including design and development), testing and evaluation, and system packaging that the MOD needs to be able to prevent critical elements of its CT capability from transferring overseas. It is critical to the UK's ability to conduct CT operations, for instance, that **design authority for the majority of its specialist equipment is retained within UK industry.**

Overview of current global and UK market

B10.12   We believe that the UK industry is generally robust in the CT arena (an assessment that covers both those companies supplying what they know to be specialist CT equipment and those whose products may have a number of uses, some of which happen to be CT in nature). Procurements are spread across many hundreds of suppliers, ranging from large multi-national companies to small manufacturers of niche devices, although in the latter case there are believed to be no suppliers who are entirely dependent on the MOD for their commercial viability. There is a system for classifying the suppliers that make up the CT sector, on the basis of their security clearance and utility of their products, known as 'List X'.

B10.13   Approximately 50% of CT-related procurement within the MOD is commercial-off-the-shelf (COTS), where existing equipment is either used directly by UK SF or others, or (in a small percentage of cases) modified following purchase to render it more covert, more robust, more usable or more effective. Whilst there are ways of improving the relationship between MOD and industry, we believe there is no urgent remedial action required to sustain key industrial capabilities.

Overt surveillance camera deployed in extreme weather conditions.

Sustainment strategy

B10.14   There is a generally recognised need within the CT community for greater early engagement with industry. At the same time, it must be acknowledged that CT covers a wide area, and a variety of engagement mechanisms appropriate to the challenges facing the different technological areas need to be established.

B10.15   The key issue preventing increased and earlier engagement of industry with the CT procurement organisations is the lack of common understanding between MOD and the CT industry. On the one hand, MOD does not have perfect knowledge of all companies that might

be able to offer solutions to its problems, while on the other hand UK industry does not have visibility of the set of CT-related capability gaps that the MOD is attempting to fill. Mitigation activities which are either under way, or are planned, to ameliorate this situation include:

- creating an Industry Liaison Officer within the Special Projects IPTs, to act as the initial point of contact for queries and approaches;

- arranging early briefing for industry on forthcoming CT-related procurement projects via Industrial Briefing Days;

- maintaining a list of 'key trusted suppliers' who could be given greater access to the MOD's CAP for the CT area, thus facilitating better transition from MOD research into commercial/bespoke products;

- maintaining an ongoing 'industry watch' to identify companies whose products might fill known capability gaps;

- using the CAP and the 'industry watch' jointly to engage early with industry and encourage them to develop, on a 'no-commitment' basis, modified off-the-shelf versions of COTS products;

- declassifying (to the greatest extent possible) Invitations to Tender (ITTs) issued in the DPA Contracts Bulletin;

- assisting companies who might have something to offer in the CT procurement arena to achieve List X status independent of an actual contract competition, thus allowing them greater access to classified requirements information;

- combining capability gap analysis and subsequent equipment procurement with Other Government Departments in order to expedite development, achieve interoperability and deliver economies of scale.

Conclusion

B10.16   The MOD approach to CT-related procurement can be characterised by several broad statements, as follows:

- it covers a diverse and wide technology base;

- it is fast-moving and reactive to known threats;

- it involves a large number of small companies and a relatively small number of large companies;

- it involves a significant amount of COTS purchasing;

- many of the requirements are highly classified (i.e. SECRET or above).

B10.17   With the exception of various niche capabilities, retention or otherwise of CT-related industrial capability within the UK occurs mainly in the areas of system design and engineering, and testing and evaluating of specialist equipment.

B10.18   No sustainability issues are currently evident, but through greater engagement with industry the MOD will be able to anticipate and address any emerging issues.

B11.1   The analysis by major platform sector in this strategy has identified a number of industrial capabilities which the UK will need to retain on-shore. Key technologies that underpin these sectors have also been discussed. Our recent Technology Strategy[1] and the National Defence Industry Technology Strategy[2] also provide specific information on the importance of technologies for defence capability. This section draws together the critical underpinning and cross cutting technologies that need to be sustained in the UK in support of sectoral strategies.

B11.2   Some of this technology will be available from the broader industrial base, for instance COTS solutions. Furthermore as the global investment in R&T continues to increase, and as an ever larger number of countries contribute to this overall growth, it is impossible for us to support cutting edge activity across all areas of R&T of relevance to defence.

B11.3   To take forward the DIS we need to identify what we need to be good at to protect our security and sovereignty. This should help to sustain and indeed enhance the competitiveness of our Defence industry. We also need an objective understanding of what we are good at and those key areas where Defence must still lead.

B11.4   Our increased collection of and reliance upon information derived from many different sensors, based on varied platforms, has many implications. These include the need for more R&T on the human factors that determine information assimilation and appropriate action, and the need to design information networks across different components of military capability.

B11.5   The life expectancy of major platforms is increasing due in part to the costs of replacement. This generates an increased emphasis on the need for equipment and systems to be flexible to meet unpredictable demands, adaptable to ensure connectivity in a network enabled world, and capable of continuous upgrade and rapid technology insertion.

B11.6   In order to address these issues, the UK will need to have a leading edge understanding in the areas of:

- overall equipment design and integration;
- design and performance of sub-systems;
- properties and limitations of key components;
- through-life capability management;
- delivering cost effective military solutions.

B11.7   Furthermore, to support future adaptation and integration our supplier base will need to have:

- the ability to control and manage equipment design, modification and integration;
- sufficient knowledge of, and access to, sub-system design and manufacture to allow modification and re-configuration;
- knowledge of, and access to, the key components and technologies needed to support upgrades.

---

[1] MOD Technology Strategy, Priorities for Defence Research [UK Restricted] dated May 2005.
[2] National Defence Industry Technology Strategy 2004.

Technologies on which the UK needs to sustain or develop technological strengths

B11.8   In order to support the industrial capabilities identified across the sectoral analysis there are a number of areas in which the UK must sustain existing technological strengths or where we should consider developing our expertise. The technologies described here support a number of sectors and capabilities. Other important technologies will be needed on-shore to support more specific sectoral or capability needs.

B11.9   These technical areas can be discussed in terms of:

- technology that can be inserted into future capability solutions and will directly improve the delivery of military effect;

- technology which will enhance enabling processes to the delivery of capability and enhance decision making;

- technology for which no specified capability or effect has been identified, but where technology watch is required as future exploitation or mitigation of that technology may become important.

Technologies for future capability solutions

B11.10   **Secure and robust communication technologies** – future defence capability against diverse and often asymmetric threats will depend on secure and robust communications. While we will benefit greatly from the large civil investment in communications, defence communications systems have to be rapidly deployed in widely varying circumstances, work in very difficult environments and often offer a high degree of security, both short and long term. Key technologies are:

- information infrastructure;
- cryptography.

B11.11   **Data and information technologies** – an increasing reliance on information and intelligence to maintain superiority over our opponents means that the ability to process, manage and exploit the wide range and large volume of data available to the armed forces remains vital to the UK. Again much of the underpinning technological developments will come from civil investment but UK defence will need to maintain and develop expertise in:

- image analysis;
- target identification and tracking algorithms;
- data fusion;
- network design and stability.

B11.12   **Sensor technologies** – sensors remain absolutely critical to nearly all aspects of defence capability including situational awareness and detection and identification of targets. Our ability to find, identify and localise targets, many of which may be heavily concealed, in cluttered environments, or mobile, relies to a very large extent on the performance and correct use of advanced sensors. Battlespace and situational awareness, including battle damage assessment, rely on the development of systems to ensure full integration between the sensors and mission and weapon programming activities. In addition technology areas such as the detection of chemical and biological agents are becoming increasingly important to our defence capability. For electro-magnetic sensing, the all-weather performance offered by sensors working at Radio Frequency (RF) (e.g. radar) is complementary to the high resolution images offered by compact Electro-Optic (EO) sensing. Where size is less of an issue, the benefits of Synthetic Aperture Radar (SAR) are slowly emerging. Consequently, technologies supporting these, as well as other more specialised sensors such as sonar and chemical and biological agent detectors, are critical to UK defence. Key technologies are:

- radar and RF engineering (e.g. phased arrays; low mass/power consumption technologies, multi-function RF systems);
- EO sensors;
- sonar;
- detectors for chemical and biological agents;
- sensor integration.

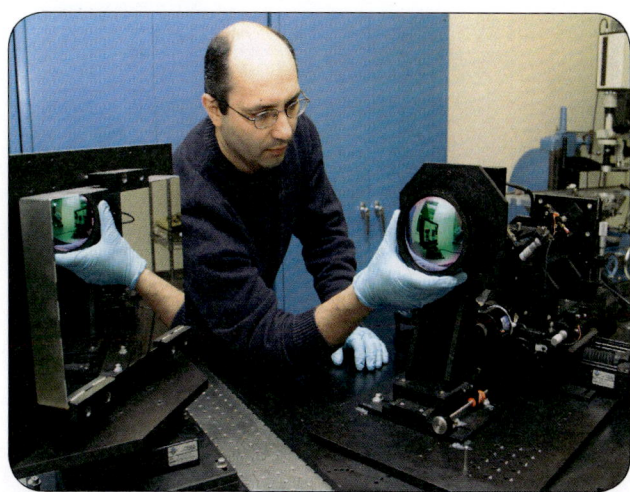

Optics lab measuring MTF of an infrared lens © BAE Systems.

B11.13 Guidance and control technologies – precision effects are central to UK defence aims and needs. These will continue to be underpinned by expertise in technologies for precision guidance and control. An area where UK has, and should maintain, a world lead is in autonomous and semi-autonomous systems for advanced guidance of weapons and uninhabited platforms.

B11.14 Electronic combat technologies – like sensors, electronic combat techniques are important across many areas of defence capability and much of the underlying technology base is shared with radar. It is likely that the UK will increasingly wish to develop and deploy integrated systems that combine sensing and electronic combat technologies in a single more efficient and effective payload. Technologies essential to this area are:

- threat detection, identification and localisation (Electronic Support Measures);
- electronic Counter Measures;
- electronic Warfare;
- RF and EO Directed Energy Weapons;
- RF engineering to support electronic combat.

B11.15 Integrated survivability – a high level of survivability, for equipment and personnel, is central to UK defence policy. In order to survive against the broad range of potential threats it is important that survivability is addressed at a holistic system level. Key technologies for survivability are:

- low observables and signature control;
- lightweight and novel armour systems;
- defensive aids systems, including EO and RF Counter Measures;
- Chemical Biological Radiological and Nuclear protection.

Signature management © Ultra.

B11.16 Technologies for remote and autonomous operation - the personnel of our armed forces are our most valuable resource and we seek to deploy autonomous systems wherever possible and appropriate, in order to protect and make best use of our people. Automation and remote operation are very important technical capabilities for delivering this aspect of policy. A great deal will be available from the civil sector but many aspects of remote operation will need to be tailored to defence needs. In addition to the guidance technologies discussed above, the main areas of interest include:

- semi-autonomous sensing and processing;
- computer assisted decision making;
- accurate underwater navigation;
- long endurance propulsion techniques;
- low power electronics.

B11.17 Automated Information and Knowledge technologies - underpinning all of the above will be automated decision aids to ensure an appropriate and timely response, particularly against short-range high speed threats. Moving forward from the core technologies discussed in B11.11, there are a number of emerging technology areas in the information and analysis space that are developing rapidly. We must track and, where appropriate, exploit advances in these areas, such as:

- information and data management;
- data mining and information extraction;
- self adapting networks;
- data storage;
- Advanced Digital Signal Processing;
- high bandwidth secure data-links;
- high bandwidth encryption.

B11.18 Power source and supply technologies - as our defence systems, including our soldiers, become more capable through being better equipped, and we seek to deploy lighter forces more rapidly and with a lower logistic burden, power sources of all types are an increasing priority. Key technologies that we will need to understand and exploit, or modify to meet military applications, are:

- efficient motive power for vehicles and power supply for systems;
- personal power sources;
- fuel cells.

B11.19 We also recognise that integrated propulsion and power plant in UCAVs could become a critical defence capability as demand for power is driven by increasingly complex embedded electronics; and that UK excellence in propulsion provides the opportunity to gain a competitive advantage in this area of technology.

B11.20 **Human Performance** – alongside the development and exploitation of the autonomous and semi-autonomous systems discussed above, the human will be the key decision maker at every level of command. It is therefore vital that research is undertaken into human decision making, cognitive processes, and techniques to enhance elements of human performance.

Technologies for enhancing capability delivery

B11.21 **Technologies to support system integration and support** - as we face a more diverse range of threats we will need to ensure that our capability is flexible, adaptable and capable of upgrade. We will, at the same time, need to ensure that the high level of connectivity and integration required for effective Network Enabled Capability is maintained. This flexible networked capability will require development and exploitation, or intelligence in accessing and militarisation, of technologies to support:

- logistics (particularly COTS technologies e.g. asset tagging & tracking);
- advanced modelling for analysis and experimentation including effects based operations;
- numerical methods for simulation and Test & Evaluation;
- new decision support models;
- open architectures;
- architectures and design to support technology insertion;
- obsolescence management;
- advanced data-loggers and failure algorithms;
- assessment and mitigation of environmental impact.

Technologies with emerging defence relevance

B11.22 In the sections above technologies important to defence have been grouped together to highlight where they are of particular importance. There are, of course, other technologies showing promise across a range of defence applications that may have either a large impact on specific defence capabilities or a more widespread impact across many aspects of defence. These can provide both an increased threat and opportunities for improved defensive capability. Examples include:

- smart materials and structures;
- Micro Electro-mechanical Machines (MEMS) for reduced size, weight and cost;
- novel energetic materials with enhanced properties;
- supersonic and hypersonic technologies;
- biotechnology and the effect on human performance (e.g. countermeasures to complex molecules with mood altering properties);
- wideband, high power electronics;
- quantum state systems for computing and communications;
- the wide potential of nanotechnology.

Technology area
Secure and robust communication technologies
Data and information technologies
Sensor technologies
Guidance and control technologies
Electronic combat technologies
Integrated survivability
Automated Information and Knowledge technologies
Technologies for remote and autonomous operation
Power source and supply technologies
Human performance
Technologies to support system integration and support

Summary of technology areas .

Next stages of analysis of R&D priorities

B11.22 Following on from this statement of priorities, we need a further level of analysis that will form the basis of further work in 2006. This will allow us to better understand and describe the nature of these technologies, where they are in the supply chain, and the level of existing UK expertise. This analysis will enable further discussion and planning with industry on:

- those areas in which maintaining a UK capability is of vital strategic importance (here the security and viability of UK supply chain is essential);
- where we can and should collaborate in both industry and university sectors at home and abroad;
- where we can buy military sub-systems and components, but to adapt and integrate (here the supply chain must be acceptable, sustainable and secure. This requires intelligent customer expertise, including assessment of product fitness-for-purpose);
- where we can buy COTS (but must have intelligent customer expertise).

Test and Evaluation

Definition

B12.1   A Test and Evaluation (T&E) capability is a combination of facilities, equipment, people, skills and methods, which enable the demonstration, measurement and analysis of the performance of a system and the assessment of the results. This can range from testing a simple switch to evaluating complex systems, such as the performance characteristics of a warship.

Strategic overview

B12.2   T&E is vital to the development, introduction into service and through-life support of the equipment used by our Armed Forces. It contributes to a variety of activities, which reduce risk to our Armed Forces.

Assurance that systems are safe and suitable for military use

B12.3   The traditional but still relevant view of T&E relates to the testing of equipment before it enters service with the Armed Forces. These activities occur following completion of equipment development and prototyping. As we move towards an incremental acquisition system there will be a requirement for more Test and Evaluation to take place on platforms and other equipment through-life, as and when upgrade and modification work takes place. There will be a wide range of testing undertaken on the equipment, in a variety of different environments and scenarios to ensure that it meets the exacting needs of the Armed Forces. This often necessitates the need for dedicated ranges and facilities, both in the UK and abroad to test amongst other things the performance characteristics of the equipment, how it operates with other military capabilities and importantly that it is safe to use.

Design and development

B12.4   During the concept and assessment phases of the CADMID acquisition cycle, there will be a significant requirement to test new technologies and concepts in particular to ensure their feasibility for use, for both new equipment procurement and upgrades. For new technologies under development, some T&E will be undertaken via modelling in synthetic environments and laboratory analysis in order to obtain an informed view on feasibility without resorting to potentially expensive physical testing.

Decision-making

B12.5   Access to T&E facilities and capabilities are important when down-selecting equipment or technology solutions. They enable the decision maker to be aware of the variety of factors relevant to the equipment or technology, how they perform and any associated risks, enabling recommendations about new and adapted equipment or technology for procurement. We place particular focus within our procurement processes on the importance of this type of early risk reduction within equipment programmes. This helps both to ensure an understanding of expected military capability but also to facilitate greater value for money later on in the procurement process as we will have developed an understanding (and where possible mitigated) of risks associated with new technology and equipment development.

Missile Real-time Simulation Facility at Dstl .

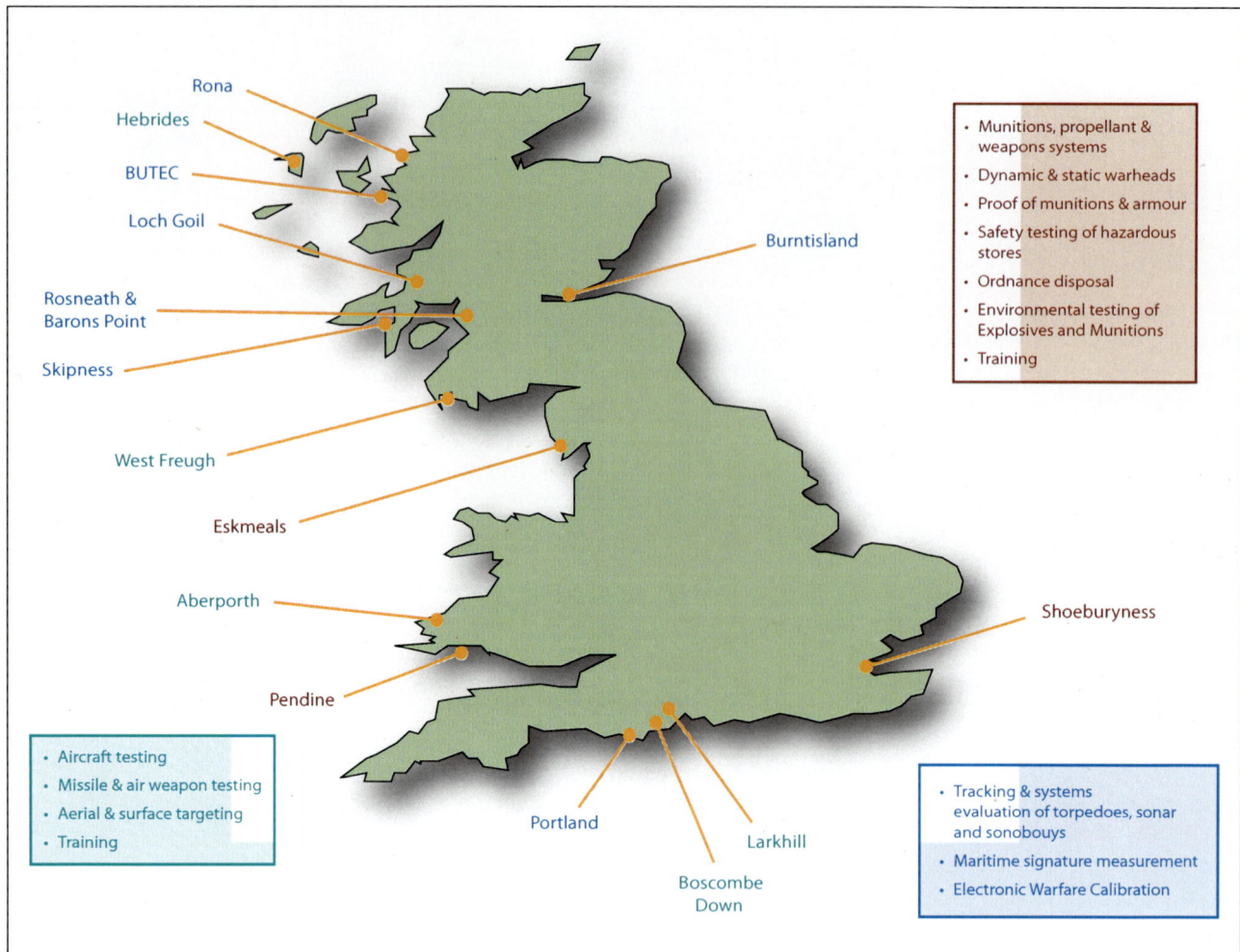

Figure B12(i) – Major LTPA T&E sites.

Tactical and operational development

B12.6   The use of T&E capabilities is paramount for testing equipment and technology within tactical and operational scenarios, in order to mirror battle conditions where a true appreciation can be gauged of the military effectiveness of the capability tested. T&E facilities also present opportunities to test the interoperability of our military capability with our allies, to help ensure effective coalition deployments and expeditions.

T&E locations

B12.7   We use a mixture of in-house, Government Owned Contractor Operated (GoCo) and commercial T&E facilities to support the acquisition and sustainment of military capability. These are a mixture of UK and overseas assets. Figure B12(i) indicates the location of the major MOD T&E sites operated on our behalf by QinetiQ under the Long Term Partnering Agreement (LTPA) and indicates the type of services they provide.

B12.8   The LTPA is a £5.6bn 25-year contract covering the MOD-owned T&E facilities previously operated by the Defence Evaluation and Research Agency (DERA). All these capabilities are kept under constant review to ensure that they continue to meet our T&E requirements and to identify potential rationalisation or efficiency opportunities.

B12.9   The map shows the breadth of T&E testing that is undertaken on-shore via the LTPA. A number of privately-owned facilities are not shown, but provide additional or similar types of T&E capability. For example, BAE Systems operate an aircraft testing facility at Warton in Lancashire, next to their fast jet production facility, similar to the testing facility operated by QineitQ at Boscombe Down.

B12.10   A number of trials are also conducted overseas because, for example, particular environmental conditions cannot be naturally replicated within the UK e.g. hot/dry (desert), hot/humid (jungle), cold/icing climates or due to other constraints (e.g. physical lack of suitable space) or because they require facilities that do not exist in the UK.

B12.11   There is a balance to be struck between retaining in the UK the required range of T&E facilities and avoiding duplication and overcapacity. This implies a clear understanding of the our T&E requirements, on which we have work in hand. It also implies the need, in the context of the LTPA, to keep under review the size and shape of the T&E industrial base. An in depth study of LTPA requirements is due to complete next year, as part of the five yearly LTPA process.

T&E vision

B12.12   The vision is to establish an effective, cost-efficient and coherent approach to allow for the testing and evaluation of equipment to support military capability both through research prior to, and then throughout the CADMID cycle. The vision includes aspects such as training, tactics, doctrine and procedures and the use of more operationally realistic and measurable T&E environments. There is little doubt that future T&E capabilities will need to provide the means to identify, and then reduce, technical, programme and cost risks from the earliest possible stages of acquisition through both synthetic and physical means.

B12.13   The realisation of our vision will require optimisation, and development, of existing UK T&E facilities coupled with analysis of other opportunities. For example:

- greater mobility, and deployability, of complex T&E equipment;
- networking of T&E and other facilities;
- greater use of modelling and simulation;
- greater co-operation with overseas Governments and industry.

Artificial T&E

B12.14   Laboratory analysis, experimentation, simulation and modelling are playing an increasingly important role in T&E activities. For example, the NITEworks experimentation capability will drive improvements in the way MOD and industry work together to develop military capability. Whilst the increasing reliance on these activities will undoubtedly reduce the need to conduct physical equipment and system testing it is not necessarily a complete panacea. There may be a shift in the balance between laboratory and physical testing, however specialist and dedicated T&E ranges, facilities and supporting personnel will still be required. The challenge is to ensure that the optimum mix is delivered and, more importantly, sustained.

T&E next steps

B12.15   Work is already underway to capture our long-term needs to support future defence acquisitions. This includes work to review both our Air T&E capabilities, routine five-yearly review of the LTPA capability requirements, and also includes an initiative, started by the European Defence Agency (EDA) and supported by national representatives, to analyse and propose 'rationalisation' of the European Defence T&E Base. The UK, amongst other nations, has been at the forefront of this process and is keen to include Government and commercial T&E capabilities into the analysis.

B12.16   Where our analysis identifies areas of T&E duplication, principally within the UK but also in the international arena, we will need to work with industry to understand the reasons for duplication and if necessary undertake any relevant rationalisation to ensure continued value for money for the taxpayer, whilst maintaining defence capability.

In July 2000 NITEworks was established to provide an integration and experimental environment to assess the benefits of Network Enabled Capability (NEC) and the options for its effective and timely delivery. NITEworks is one of the key means of carrying NEC from ideas to delivery. This is achieved through experimentation in a simulated battlespace to identify the benefits of NEC and the practical steps we can take to deliver improved capability to the front line. NITEworks operates as a unique MOD-industry partnership which sees customers and suppliers working together to realise the potential of NEC.

Conclusion

B12.17   In some cases a UK based T&E capability is essential for, amongst other things, certain quality assurance, safety or operational security needs and sovereignty of access. In other cases the important element is to retain the ability to direct, understand, analyse and verify T&E results rather than actually conduct testing on-shore, subject to certain safeguards including security of supply.

B12.18   We will work with industry to identify where such distinctions can be safely made based on the principle that facilities are retained for defence capability purposes. The current strategic intent in the medium term is to retain T&E capability within the UK, but look for overseas co-operation where appropriate. The EDA work may lead, in due course, to a longer-term strategy to consolidate T&E capabilities across Europe.

B12

Test and Evaluation

section C

Taking forward the DIS - the challenges for change
Getting down to work - putting the DIS into action

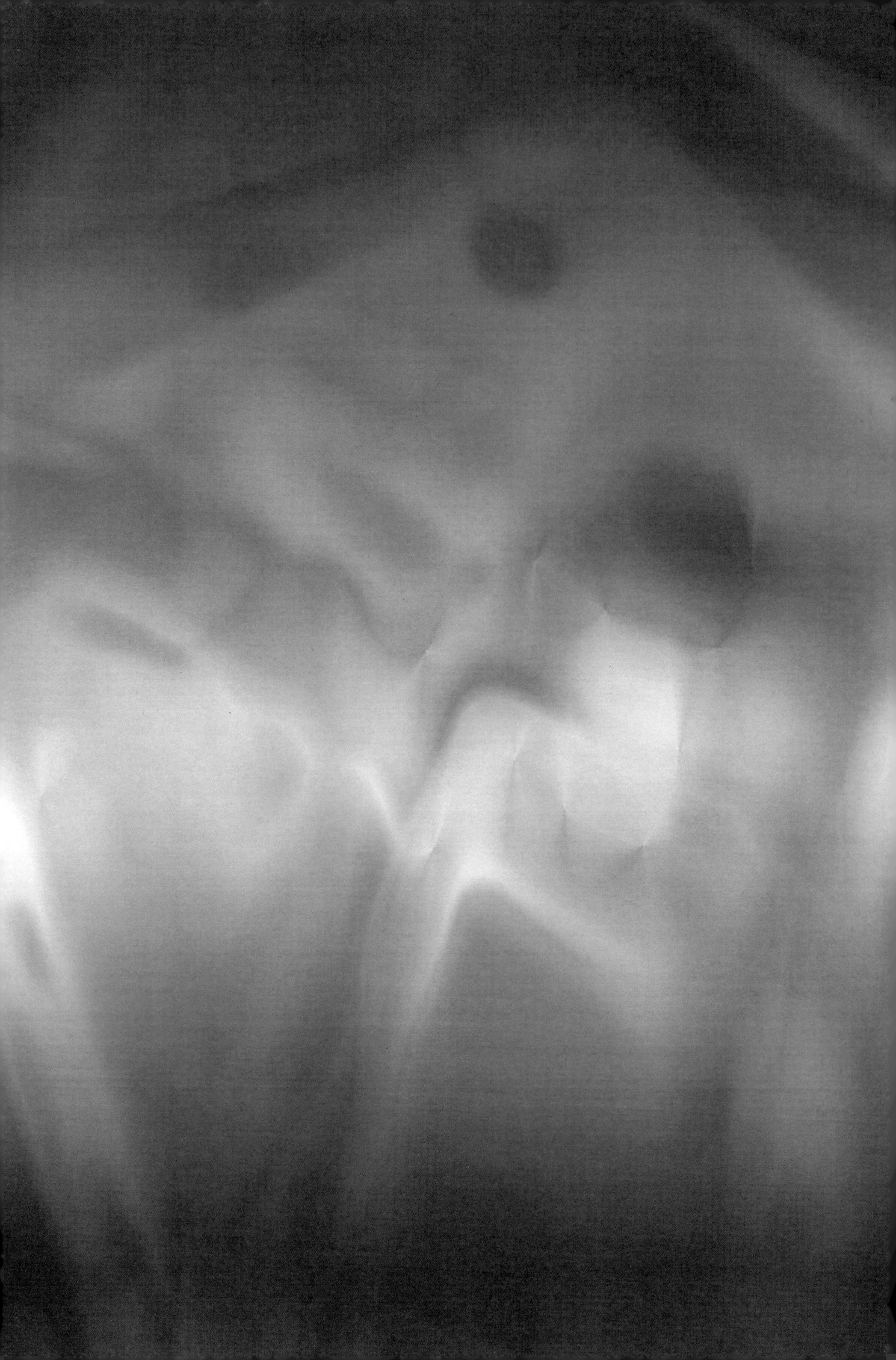

C1.1   The DIS presents real and urgent challenges to the way in which the Department conducts the business of defence acquisition in future. If we are to step up to those challenges and translate the strategic intent into a reality, we must acknowledge the need to change. The strategy will not deliver unless the defence acquisition community as a whole, including industry as well as MOD, make the essential transformation in our behaviours, organisations and business processes.

C1.2   Our Smart Acquisition initiative has delivered significant benefits for Defence in the form of improved performance and delivery of affordable, battle winning capability to the Front Line. The basic principles of Smart Acquisition still hold true and existing change programmes throughout the Department provide a solid foundation on which to build.

'There has been further progress on measures to improve performance within the Defence Procurement Agency and elsewhere in the Department. These improvements focus on the following areas: performance of key suppliers; the skills and development of staff; project and risk management; increased use of trade-offs between time, cost and capability of equipment; better joint working of those responsible for acquisition within the Department; and stronger project scrutiny at all levels'.

National Audit Office Report – Major P rojects Report 2005, HC595-1, Session 2005-06

C1.3   We must address the remaining challenges across the Department of embedding a through-life systems approach, achieving a better and more transparent relationship with industry, improving

risk and performance management, and tailoring our procurement approach to the needs of individual acquisition programmes.

C1.4   The nature of acquisition is evolving and we face an increasingly demanding and complex environment. Closer collaborative engagement between us and our industrial suppliers will be vital if we are to continue to deliver the improvements that the Armed Forces and UK taxpayers demand. The increasing pace of technological change, linked to a demand for delivery of projects that combine new equipment with other elements such as through-life support and training as an integrated capability present challenges that both the Department and industry must face together.

C1.5   Our future approach to acquisition will therefore be built around the objective of achieving:

- primacy of through life considerations.
- coherence of defence spend across research development, procurement and support.
- successful management of acquisition at the departmental level.

C1.6   Taken together these objectives will form the basis of our acquisition reform programme, and our response to the challenges of the DIS. We intend to build on our achievements to date - consolidating success and embedding best practice – but also recognising that we need to drive reform where it is needed.

Our Values for Defence Acquisition

C1.7   Successful acquisition depends not just on getting our strategy, organisation and management processes right. We must ensure the fundamental enablers are right, such as the cultural environment in which we do business, the values and behaviours to which we adhere.

A meeting of the National Defence Industries Council.

Defence Values for Acquisition

Defending the United Kingdom and its interests

Strengthening international peace and stability

A FORCE FOR GOOD IN THE WORLD

We achieve this aim by working together on our core task to produce battle-winning people and equipment that are:

- fit for the challenge of today;
- ready for the tasks of tomorrow;
- capable of building for the future;

Our defence values for acquisition

By working together across all the Lines of Development, we will deliver the right equipment and services fit for the purpose required by the customer, at the right time and the right cost.

In delivering this Vision in Acquisition, we all must:

- recognise that **people are the key to our success**; equip them with the right skills, experience and professional qualifications;
- recognise the **best can be the enemy of the very good**; distinguish between must have, desirable, and nice to have if affordable;
- identify **trade offs** between performance, time and cost; cases for additional resources must offer realistic alternative solutions;
- **never assume additional resources will be available**; cost growth on one project can only mean less for others and for the front line;
- **understand that time matters**; slippage costs – through running on legacy equipment, extended project timescales, and damage to our reputation;
- **think incrementally**; seek out agile solutions with open architecture which permit "plug and play"; allow space for innovation and the application of best practice;
- **quantify risk and reduce it by placing it where it can be managed most effectively**; stopping a project before Main Gate can be a sign of maturity;
- **recognise and respect the contribution made by industry**; seek to share objectives, risks and rewards while recognising that different drivers apply;
- **value openness and transparency**; share future plans and priorities wherever possible to encourage focused investment and avoid wasted effort;
- **embed a through life culture** in all planning and decision making;
- **value objectivity** based on clear evidence rather than advocacy; ensure that we capture past experience and allow it to shape our future behaviour;
- realise that **success and failure matter**; we will hold people to account for their performance.

C1.8 In October 2005, we announced a core set of defence values for acquisition which build on our Defence Vision to shape the behaviour of all those involved in acquisition, including: Ministers, the Defence Management Board, customers at all levels, the scrutiny community, project teams in the delivery organisations and private sector partners.

C1.9 We will **embed these values** for acquisition in our partnering arrangements through clear leadership at all levels by:

- ensuring all key acquisition decisions refer explicitly to how they reinforce and demonstrate these values;
- reflecting these values in our acquisition personnel objective setting and reward structures;
- embedding these values in our core acquisition guidance documents and education programmes; and
- using these values as the basis of the way we develop our future relationship with industry.

Achieving the gold standard

C1.10 The recent Value For Money report from the National Audit Office[1] identified several examples of 'Gold Standard' performance in our projects "**with a number at the very forefront of good project control**", against a benchmark of worldwide best practice in other sectors. In setting out our agenda for change, we must take the whole of the Department forward on a broad front, embedding best practice in everything we do. We set out below our priority areas for action and the steps we will take to achieve the changes we are seeking.

Through life relationships with industry

C1.11 Defence acquisition must be able to adapt to the increasing uncertainty in our external environment and the future operational

[1] National Audit Office 'Driving the Successful Delivery of Major Defence Projects: Effective Project Control is a Key Factor in Successful Projects'. HC 30 Session 2005-2006

requirements; programmes that are increasingly complex, of higher value and higher risk; a greater drive for innovation and continuing cost improvement, and models of product and service delivery that are more through-life and long term in nature. Relationships between Department and industry that are purely transactional and conducted at arms-length will struggle to meet these challenges. Increasingly they demand the use of a different style of relationship.

'Successful working relationships are characterised by soft factors such as team working, trust and honesty. When the Department and its industry partners on a project display these behaviours they are more likely to develop a common understanding of the task, the progress being made and give early warning of problems. When a project operates in a supportive and open corporate environment the other parts of the project's own organisation, such as senior management, are more likely to have timely and accurate information about the status of the project to enable them to make sensible decisions.'

National Audit Office – Driving the successful delivery of major projects, HC30 Session 2005-06

C1.12   The emphasis on our future approach to ensuring value for money has highlighted the need to place greater emphasis on fostering better, and where appropriate, longer term relationships with our key suppliers, and the use of appropriate commercial tools, including competition of formal parnering agreements. This must be underpinned by greater openness and transparency, with a common and more explicit understanding of how to achieve best value for both Defence and industry. While there are examples of good practice in fostering good relationships, including in Defence Estates, these are not as widespread as they should be.

C1.13   'Partnering' defines how the parties conduct themselves and the working attitudes that are valued. While it can describe a legal framework of partnership, the basic ethos and the associated behaviours are not restricted to any particular method of contracting.

C1.14   Learning the requisite behaviours and skills, as well as other professional procurement competencies, takes time and experience. We recognise that it will be some time before all our acquisition specialists are able to demonstrate significant experience of practising these partnering behaviours. We are committed to retraining and developing our people to develop the competencies required, and will encourage individuals to plan their careers with a view to preparing them for such positions. Nevertheless, we recognise that in the short term we may not be able to find internally all the individuals we require who have or can develop quickly the requisite skills and experience, and would therefore need to recruit externally. To ensure that we maintain a vibrant, highly skilled acquisition community in the department in the longer term we will also, in line with the Professional Skills in Government agenda, actively encourage continued movement between the Department and Industry.

C1.15   The business environment to which we aspire is one in which there is:

- a relationship which is less 'adversarial' in style, based on a mutual understanding of where the motivations and interests of each party lie, acknowledging and managing the areas of difference and tension;

- a willingness to share information with industry in a spirit of openness and transparency at all levels. This must begin from the earliest stages of the project lifecycle, involving industry more closely in helping us to identify and shape the requirement;

- clear boundaries of responsibility and authority in the dealings between us and industry with explicit codes of practice and behaviour that are actively managed;

Working jointly with the MOD, Rolls-Royce Defence Aerospace has achieved improved mission availability whilst reducing cost of ownership with its MRMS - Mission Ready Management Solutions- strategy. This focuses Rolls-Royce's aftermarket support on the creation of cost-effective solutions, specifically customised to the requirements of individual platforms and customers and designed to provide enhanced value through the application of Rolls-Royce's IPR and unique product knowledge. Rolls-Royce and the DLO have jointly embarked upon a partnering approach to transform the logistic support to the Front Line Commands.

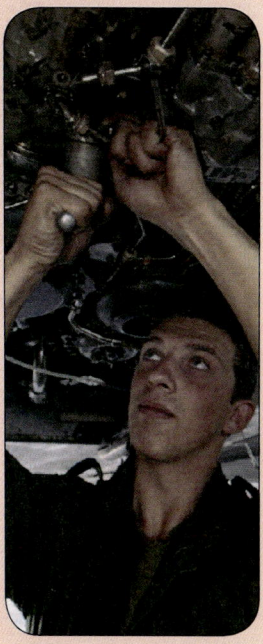

Whilst MRMS is a relatively new concept, exemplified by the benefits flowing from the support solution for the RB199 engine in the RAF's Tornado aircraft, which in a pilot contract delivered 30% cash savings whilst improving engine availability, it builds on a track record of previous successful innovation. The successful Total Support Programme for the Spey 250 engine in Nimrod R1/MR2 led the way. Under this arrangement, average engine on-wing time has more than doubled, unplanned rejections have been reduced by 50% and spares consumption has reduced by over 20%.

The achievement of these innovative support solutions has been based on the establishment of a joint Rolls-Royce/DLO team, working within a partnering arrangement. The arrangement included developing a jointly owned through-life cost model, shared assumptions database, joint risk register and an integrated cost reduction programme involving Rolls-Royce, its supply chain, the DLO and Strike Command.

By first clearly understanding the desired outcomes of all parties, a contracting model was developed which has established effective performance-based incentives for reducing the through-life cost of RB199, not least by applying Lean Support Principles to maximise total support effectiveness.

This innovative approach has relied on the adoption of entirely different behaviours by industry and the MOD, which have been characterised by a change in leadership style and co-location to achieve successful joint team working, based on shared information and aligned business objectives.

- an ethos that encourages potential problems to be brought to light early, with effective mechanisms for the timely resolution;

- more widespread use of 'partnering arrangements'[2] in our contracting relationships where appropriate; having explicit codes of practice and behaviour that are actively managed;

- relationships between MOD, industry and others that encourage innovation and facilitate the insertion of new technologies and upgrades into military capabilities. In some areas this will require greater understanding and management of the critical points in the industrial supply chain.

C1.16   We will therefore:

- place a greater **emphasis on joint team behaviours and relationship management as part of the core business of all our acquisition**. This will include developing project teams which have the tools, skills and expertise to facilitate more effective relationships;

- **create a new Defence Commercial Director post in the centre of the Department with a pan-Defence strategic outlook.** This appointment has already been advertised. The post holder will be responsible for driving forward the commercial aspects of the Defence Industrial Strategy and, in particular, for developing partnering arrangements with industry that embed the right behaviours and incentives for both parties;

- **place greater emphasis on partnering behaviours,** the importance of partnering arrangements, and the ability to foster effective relationships, **in our supplier selection decisions**;

- invest **increased resource and commitment into driving forward the Key Supplier Management initiative** to deliver a framework that will improve and develop our knowledge, understanding, and relationships with key suppliers throughout the supply chain focusing on mutually beneficial improvements in decision making and performance, while also improving risk management across our portfolio;

- **improve commercial awareness and the understanding of industry for our acquisition staff at all levels**. This will include a greater use of joint training and development, and creating more opportunities for short, focused exchanges of staff between the Department and industry;

- **ensure greater collaboration** between defence, industry and the universities in the fields of science, technology and engineering. We will develop the supplier base by building on our existing plans to compete more of the research programme and forming partnerships between government, industry and universities;

- **publish a guide, intended for an industry audience**, making it clear whom within our acquisition organisations has the lead responsibility and can speak authoritatively on particular subjects. This guide will be published electronically and will be updated through the Acquisition Management System.

Defence Estates Supplier Association

Defence Estates (DE), through the adoption of Prime Contracting, has generated a major rationalisation of the traditional property management and core works contracts, reducing them from over 200 into 5 regional based medium term contracts of between 7 and 10 years duration. A significant proportion of DE's business is or will be contracted out to a much-reduced number of suppliers for the medium to long term.

As part of DE's Supplier Management strategy we aim to improve the overall working relationship with key suppliers following this move to longer, high value contracts. We believe that the Supplier Association format provides us with a formal structure through which we can achieve this objective. The associations are intended to be inclusive and a 'joint' initiative, therefore cannot work without the commitment of DE's key suppliers.

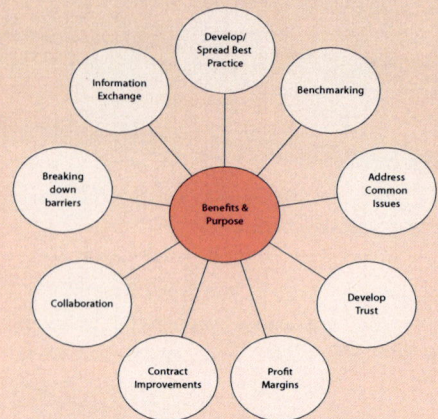

DE organised workshops with its suppliers during November 2005 to launch the Supplier Associations and are confident that by early 2006 the associations will be fully established and will have identified specific areas on which DE can work collaboratively with its suppliers, in order to generate tangible outputs and efficiencies.

Delivery of integrated solutions

C1.17   We will continue to promote adoption of a more integrated approach to the delivery of military capability, through our focus on Through Life Capability Management as the basis of all our acquisition activities. While the principle of an empowered equipment capability customer is now well embedded into our acquisition system, we need to improve the synchronisation of other contributing defence Lines of Development (see text box). The delivery of new and enhanced military capability requires orchestrated action across complex change programmes in addition to the equipment itself.

---

[2] A 'partnering arrangement' is not generally a legally binding form and it can be applied to any contractual relationship. It differs from a formal 'Partnering Agreement' in which the MOD and a supplier form a legally binding, collaborative entity.

Defence Lines of Development

- Training
- Equipment
- People
- Information
- Doctrine and Concepts
- Organisation
- Infrastructure
- Logistics

The **Future Rotorcraft** (FRC) programme, initiated in 2004, combined the funding and requirements of several helicopter equipment projects to achieve increasing coherence and cost-effectiveness within capital and support cost constraints.

The programme needed to balance the priorities for investment across the non-combat (e.g. Search And Rescue) and combat rotorcraft capability domains of Lift, Find and Attack in both the Land and Maritime environments. This was driven by the need to address ageing and obsolescence issues on elements of our current fleet; all three Services have rotorcraft responsibilities.

A **Senior Responsible Owner** was appointed to provide leadership within the Department for this complex challenge. Additionally, one of the three Directors of Equipment Capability with responsibility for rotorcraft was vested with the **Single Point of Accountability** for all rotorcraft programmes. The third structural change within the Department was the appointment of a dedicated FRC IPT Leader to marshal the efforts of IPTs in this key capability area.

C1.18   We have created a number of through-life IPTs with dual accountability to the DPA and the DLO, to provide a cradle to grave approach to equipment management. The DLO is already driving an end to end through-life view of logistic support solutions that provides opportunities and incentives for industry to align with our capability needs.

C1.19   There is much more that we still need to do, if we are to:

- get the best value for money from the defence industrial base in the delivery of battle-winning capability to the front line, not just at the level of individual projects but across multiple projects and programmes;

- manage the inter-dependencies between our projects more intelligently and effectively; and

- exploit opportunities to provide more cost effective ways to provide military capability through innovation and change in the non-equipment Lines of Development.

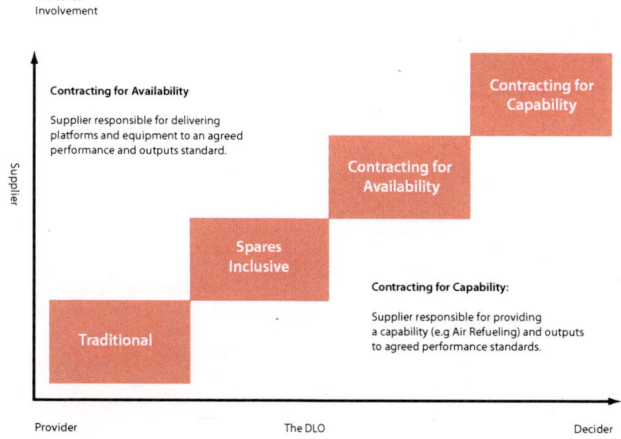

Figure C1(i).

C1.20   We will take action to **create a strong programme management environment** around our projects that will:

- manage the **overarching portfolio of projects within a capability area**, including research and technology, capability upgrade and in-service capability in a coherent manner. Programme teams will be accountable for the initiation and execution of projects, working with suppliers to reduce the likelihood of individual projects over-running and the impact on the wider acquisition budget if this is unavoidable;

- manage **cross project issues**; oversee the integration of projects and other Lines of Development into military capability; increase our capability to trade-off between performance, time and cost; and provide a focal point for underpinning industrial base issues and ensure coherent engagement with the market. This will include an intelligent approach to the structuring of the supply chain to maximise innovation and nurture the necessary systems engineering capabilities;

- act as **overarching design authority for the capability area**; to enable faster and cheaper capability upgrade; manage the insertion of new technologies and increments; better focus on the pull-through of new technology; exploit opportunities to reduce through-life costs by more coherent management of the total portfolio of equipment and projects;

- invest in **developing programme management capabilities and competence** within acquisition. This will ensure the availability of the key enablers defined by the OGC for Successful Delivery Skills and Centres of Excellence.

C1.21   We will run **Pathfinder programmes** to test and de-risk this approach to capability programmes with close involvement from industry in the areas concerned.

C1

Taking forward the DIS - the challenges for change

Defence Industrial Strategy   **135**

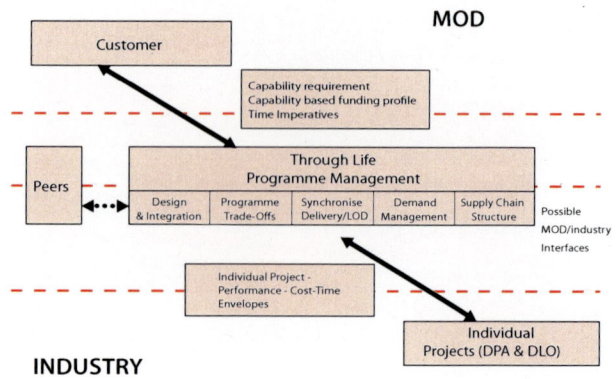

MOD

INDUSTRY

Figure C1(ii).

Pathfinder Programmes

Sustained Maritime Surface Combatant Capability. The long-term sustainment of the capabilities currently delivered by Maritime Surface Combatants alongside a solution for the Key User Requirements previously identified by the Future Surface Combatant programme.

Sustained Armoured Vehicle Capability. The long-term sustainment of the capability delivered by the current and programmed Armoured Vehicles.

C1.22   The key features of the Pathfinder programmes will be:

● culture change through setting the right values and behaviours;

● a programme approach to through-life capability management;

● effective integration across all the Defence Lines of Development to deliver sustainable military capability;

● effective techniques for capability trade-off with early industry engagement in capability analysis;

● Defence/industry joint working to understand and address the dynamics of the supply chain and sources of innovation.

C1.23   We are working up business cases for each of the Pathfinders with the intention that both programmes will launch in the first half of 2006. We also intend to capture lessons emerging from our FRC, SAM and Acquisition for NEC programmes as a means of identifying best practice in programme management.

C1.24   Work is also underway to assess the application of programme management approaches to the departmental Information Systems portfolio, and to consider the need for clear end-to-end ownership of the processes used for the delivery of the Command Information Systems within MOD.

Innovation, agility and flexibility

C1.25   We must be able to respond to the rapidly changing strategic and operational environment by adapting current and future capabilities exploiting the opportunities offered by technology innovation. We must remain alive to developments in the commercial market, particularly in the fields of information and communication technologies that are evolving at a pace that can outstrip the ability of our procurement processes to respond.

C1.26   There is also much that we can learn from our handling of Urgent Operational Requirements, recognising the unique circumstances in which they are generated and delivered. Nevertheless, such rapid procurements are a testament to the dedication and ability of our acquisition community in times of crisis, and show a pragmatism and preparedness to trade performance and make use of innovative contracting approaches.

C1.27   Alternative acquisition lifecycles, such as incremental acquisition, are better suited to these challenges and allow for a staged approach to the reduction and management of project risk. They also offer flexibility to enable the insertion of new technology and a rapid response to evolving requirements and operational circumstances.

Faced with high levels of programme risk and a rapidly developing technology base, the **PICASSO** project adopted an incremental strategy, combined with the appointment of a Prime Systems Integrator (PSI) and a Joint Project Office.

The strategy has enabled the PICASSO team to deliver capability in manageable programme increments, with assessment activity conducted concurrently with the delivery of each increment. The PSI is responsible for assessment, delivery and integration of future increments and the through-life Operation & Maintenance of the current and future capability. This approach has been effective in mitigating the programme risk by giving the team the agility to react to emerging issues and to trade performance, time and cost for each increment based on changing requirements.

It has allowed MOD and industry to pool ideas and has promoted a clearer understanding of how industry solutions can meet MOD's needs. This approach also gives MOD the opportunity to direct industry to integrate known third party solutions into legacy and new capability. Progress to date has been excellent and the project is meeting its 10% confidence level estimates.

C1.28   To inform and support such alternative acquisition lifecycles we will require new models and analysis that enhance our ability to:

● understand how capability and technology combine to deliver us the military effect;

● develop clear options for consideration;

● conduct balance of investment to inform decisions between investment in platforms and weapons or enabling systems.

C1.29 We must also look to research and technology (R&T) to provide more affordable and cost-effective capability. This must provide a greater focus for our future programmes. Examples of where R&T programmes have helped to address affordability issues include:

- seeker research work on the Brimstone air-launched anti-armour missile allowed savings of £15-20 million, and increased the capability.

- Flight control research for the JSF, providing estimated cost saving of £40 million for fixed costs and £3.6 billion over the next 30 years for a research cost of only £50 million.

The Tornado F3 Sustainment Programme (FSP) is a collaborative project to upgrade the Tornado F3 to fit AMRAAM and ASRAAM missiles and to make improvements to the radar and mission computer software. The project combines several upgrades into a single, coherent package of work, delivered in increments. The small joint team (comprising Tornado IPT, Fast Jet and Weapons Operational Evaluation Unit, BAE Systems, QinetiQ and the missile, radar and computer contractors) have adopted concurrent engineering principles, strong teamwork and simultaneous development, trials and operational evaluation.

This has reduced the projected time (by 50%) and cost (by 30%) as well as allowing front line views to be incorporated early in the development process. Already at the midpoint, the programme has proven resilient in the face of a number of technical problems but has still achieved all of its stretch milestones. This approach is being broadened into the Tornado Capability Development and Sustainment Service to accelerate the introduction of new technology onto the Tornado GR4 from 2010 to 2007; this opens up the architecture of the aircraft to make future technology insertion easier and hence reduce whole-life costs.

C1.30 We will seek to realise the value of innovation by exploiting it better in new equipment or new processes, including a greater emphasis on the development of demonstrators[3]. These offer major benefits by providing a means to challenge conventional views on system solutions and offering early risk reduction. We will also improve the planning of research exploitation, and to use this to inform funding decisions. This will help the MOD and industry allocate funding and align equipment and the supporting R&T programmes.

[3] *The Management of Defence Research and Technology Part 4 NAO Report March 2004*

C1.31 Adoption of open system architectural principles can assist in developing modular solutions that maximise the opportunities for technology insertion, as well as promoting innovation and competition within the supply chain. These principles ensure the equipment design process employs declared, common standards, interfaces and supporting formats.

C1.32 There is further scope to exploit synthetic environment approaches and the use of experimentation techniques involving the end user. These techniques can also be used to capture and refine requirements; identify and de-risk the final solution; ensure all Lines of Development have been addressed; and support an end to end perspective on the key integration and interoperability issues.

C1.33 Exploitation of e-procurement techniques offers the potential to further streamline the acquisition process, reduce bureaucracy and promote an effective collaborative working environment with industry. This applies to the procurement of commodity items, the acceleration of bidding and tendering, and support to virtual teams working with industry. E-procurement does not just replace paper transactions. It transforms business processes and fosters more effective trading partnerships with our suppliers.

Submarine Combat Systems Open Systems Architecture aims to harness the extraordinary development in COTS processor power to manage traditional military system obsolescence, reduce whole-life costs and sustain capability. It is based around three principles: improving open access to software applications from a variety of sources; reducing dependence on a single provider of bespoke military hardware; avoiding "lock in" to a single provider by establishing MOD led activities to oversee and separate application development, selection processes, standards development and management of systems integration.

Technical demonstration has shown the maturity of the approach, drawing on US experience. The process has required industry to accept the reduction in business volume and adopt OSA to remain engaged, without MOD liability for the down-sizing. Greater processor capacity and "smarter" applications considerably improve existing performance. A major benefit of the OSA is that new applications in response to urgent capability demands can be hosted more rapidly and at lower cost; major system cycle time for introducing new capability is reduced by approximately 75% and it is expected that support and upgrade costs will be reduced by approximately 40%.

C1.34   The major enabler for e-procurement is already in place. The Defence Electronic Commerce Service provides the principal portal for e-business transactions between the Department and its trading partners. Suppliers already trading through this route have identified a number of key benefits including: less paperwork and administration; quicker payment; greater competitive advantage through reduction in inventory holdings; direct supply to the customer and greater control of the supply chain.

C1.35   We will take action to **make our acquisition approach more innovative, agile and flexible at the project level, by:**

- **streamlining decision making**, recognising that delays in decision-making have a cost;

- **fully exploiting the potential of e-procurement**, throughout the MOD acquisition organisations;

- **rebalancing our governance arrangements** towards a greater emphasis on assurance that risk is progressively reduced through the life of the project;

- recognising that **being prepared to cancel projects, if necessary before Main Gate, is a necessary part of a healthy acquisition system** recognising the impact this may have on bidders;

- **increasing the tempo of procurement**; matching to the underlying speed of technology change and changing operational needs, moving to the presumption that all projects should use **more flexible approaches (such as incremental or evolutionary acquisition);**

- **systematically reviewing** the procurement strategies for our existing pre-Main Gate project population to identify potential benefits and opportunities for improved delivery through a more **flexible** approach to acquisition;

- **developing an improved approach to technology insertion**, allowing us to 'plug and play' new technologies more easily into our major systems and platforms, maintaining a capability edge and ensuring interoperability with high technology allies such as the US and France. Specifically, working with industry to identify innovative best practice and benchmark UK performance against other nations for cost effective technology insertion by the end of 2006;

- **working with industry and universities to identify national sources of innovation** and where the important technologies we need are:

- improving the **pull-through of technology from research into capabilities**; increased technology scanning to identify potential opportunities and threats, and the funding and focus of technology and capability demonstrators;

- exploiting **synthetic environment and experimentation** approaches to reduce risk and facilitate early engagement of the operational customer in system design.

Consistency in our approach

C1.36   Empowered and accountable Integrated Project Teams remain a core principle of how we conduct acquisition. But some of our project teams behave inconsistently and may fail to follow best practice. Empowerment can also result in approaches that are effective at project level, but which may not be in the best overall defence interest.

C1.37   We will take action to **create a consistent and clearly defined operating framework for how we conduct, govern and control our projects.** This framework will:

- embed the Defence Values for Acquisition throughout the organisation;

- establish a strong and professional 'doctrine' for how the Department conducts its acquisition business;

- reduce the burden of compliance and governance and improve the speed and responsiveness of our decision-making;

- be underpinned by a set of obligations that clearly articulate what customers, suppliers and other stakeholders can expect in their dealings with the acquisition organisations.

C1.38   The DPA intends to review and clarify its operating framework during 2006, working in conjunction with the other areas of acquisition to ensure consistency and alignment. The DLO Procurement Reform and Category Management initiatives have also made progress in promoting greater coherence and consistency among their logistics teams. We will work across the acquisition community as a whole to exploit opportunities for improved joint working and commonality of approach, including reviewing and streamlining the Acquisition Management System to ensure that content is relevant, authoritative and readily accessible to practitioners. Together with the Defence Values for Acquisition, and the Acquisition Handbook, this will form a coherent and readily accessible suite of guidance.

'Top team oversight of major programmes is critical to success. IPPD recommends that departments establish a centre of excellence, combining the roles of programme office and departmental capacity/ capability building. The centre of excellence will ensure Management Boards and Ministers have the systems and data they need to prioritise, monitor delivery, and balance risk against departmental capability.'

Improving Project and Programme Delivery: Office of Public Service Reform

C1.39   We will implement the OGC Commerce Project and Programme management 'Centre of Excellence' model to ensure that we:

- continue to drive toward the NAO 'Gold Standard', and toward best in class professional procurement practices in public and private sector organisations world-wide;

- fully integrate risk management into our project management approach;

- identify and progressively implement a set of best in class project control and reporting tools.

C1.40   We will also embed improved joint working between the DPA and DLO exploiting opportunities for improved alignment, commonality of approach and economy of effort. These include:

- continuing to launch through-life IPTs;

- rationalising the provision of enabling services that provide specialist skills in support of IPT activity;

- improving our knowledge management and our learning from experience across the acquisition community.

Professional delivery skills

C1.41   We must ensure that the appropriate training, development and professional standards are in place for all those involved in acquisition, and that staff receive due reward and recognition for their competence and for their achievements in project delivery.

C1.42   The Acquisition Leadership Development Scheme is key to achieving this and provides a clear career anchor and development structure for acquisition professionals. There are currently some 664 members. Senior leaders in their respective professional fields have been identified as development partners to: determine the future skills needs; develop and advise on career paths; and inspire individuals to acquire the skills the acquisition community requires.

C1.43   Current initiatives in this area include:

- a MSc in Defence Acquisition Management launched by the Defence academy;

- a Business Graduate Development Scheme providing acquisition teams with a new pool of high calibre professional commercial managers;

- graduate recruitment schemes to enhance critical skills in project management, engineering and finance.

C1.44   The Civilian Workforce Plan for 2005 describes our overall strategy for our civilian workforce and will enable us to maximise our pool of talent and match people and skills to the demands of the job. This places a high priority on programme and project management skills and on ensuring that our recruitment, reward and recognition practices deliver people in the numbers we need with the necessary skills and motivation. We will:

- **address the shortages in Project Delivery Skills** within the department through a programme to address critical shortages of project and programme management and acquisition leadership skills. This will include key milestones to be achieved by April 2006;

- improve our **recruitment, reward and recognition** practices to deliver acquisition staff of the number and quality we need and whose behaviour demonstrates their commitment to our

Acquisition Values. This will include looking to new approaches to recruitment, pay and grading, and reward, to attract, develop and retain people with the right skills. We must also ensure that our most challenging projects attract, and are led by our very best, and that we grow people for those roles through the course of their careers. Individual and team achievements in project delivery must also be appropriately recognised and rewarded. An evaluation of current initiatives will be undertaken with the Project Delivery Skills Programme and completed by October 2006;

- make increased use of **professional accreditation schemes** for engineers, project management, finance and commercial staff in our professional acquisition streams;

- co-ordinate the effort on Professional Delivery Skills with the action to **improve the Science and Technology skills base** within Defence;

- **increase investment in systems engineering** skills and training in the Department and in industry.

- **place greater emphasis on staff continuity in delivering projects**. The tenure of staff in key posts must ensure greater continuity of responsibility, and relationships across critical phases and events in the project life.

C1.45   We are also committed to seeking ways to exploit delivery and project management skills in industry. We will:

- **work more closely with industry in developing acquisition skills and professionalism**. This will include reinvigorating the joint working initiatives on human resources issues, including joint learning events, shared development opportunities, short and focused interchange opportunities for acquisition staff, and more joint education through the Defence Academy;

- **explore alternative models for independent project management** of major projects. We will pilot this approach on two acquisition programmes.

Moving forward

C1.46   This section has set out an ambitious and challenging change agenda that will require committed and visible leadership within the Department. We intend to drive hard to realise the benefits as soon as

possible, recognising that it will be some time before results are fully visible externally. This change agenda will be led and driven by the Acquisition Policy Board (APB) which will regularly review progress and ensure change activities are adequately resourced and supported across the Department.

C1.47   We have created a Directorate of Defence Acquisition responsible to the APB for coordinating this programme of change, tracking and capturing benefits, and ensuring that obstacles are recognised and addressed.

C1.48   All of this will be necessary if we are to improve our acquisition performance. But we recognise that it may not be sufficient. We need better to understand where our current processes, structures and organisation support, encourage, hinder or obstruct our ability to achieve the objectives of this change programme, and to address the obstacles of the DIS.

C1.49   We will appoint a senior official to review our current acquisition construct and make recommendations for change where needed. This individual will report directly to the Permanent Under Secretary and, through him, to the APB. They will have a clear remit to range across the whole of the Department's business and be encouraged to take a broad view of the acquisition process. They will report progress to the APB on a regular basis, with final recommendations by May 2006 for early implementation.

The challenge to industry

C1.50   The Department is committed to driving this change agenda. We will be looking for parallel commitment from industry to:

- plan more effectively and jointly for the long term, embracing the vision of through life capability management so as to focus on meeting our requirements in the most cost-effective way in whole-life terms;

- invest in growing and maintaining a high-quality systems engineering capability within the UK, at all levels in the supply chain where we need key systems and sub-systems to be designed and engineered;

- join us in our efforts to promote greater interaction and collaboration between Defence, industry and the universities to stimulate innovation in science, technology and engineering;

- embrace the use of open systems architectural principles and incremental acquisition principles throughout the supply chain, and help us to find more cost-effective approaches to technology insertion;

- work jointly to foster better understanding of each others' objectives and business processes, including a greater commitment to joint education, staff development and interchange opportunities;

- promote the use of partnering behaviours in industry's interface with the Department at all levels, so as to encourage trust, openness, transparency and communication.

---

[1] See Chapter A5, para A5.19 on developing the MOD's R&T programme.

C2.1   The Defence Industrial Strategy sets out a comprehensive agenda for change, both in how we approach and interact with the market place in several areas and in how we and industry behave and are organised. Much effort has been expended – within the MOD, across Government and in industry – in putting it together.

C2.2   But all this will be for nothing unless Government and industry are prepared to work together to address the real challenges that we face if we are both to maintain the industrial capabilities and technological skills that we require in order to maintain the operational capability of our Armed Forces in a manner that is attractive in a business sense. For instance, we and industry agree that in general we must gain a much deeper understanding of the supply chains for defence; this will require significant, ongoing work on a number of fronts. In that sense, the publication of the DIS is the start, rather than the end, of a process. Its detailed implementation is the hard work to come.

C2.3   We recognise this, and the importance of making the DIS more than simply words. For our part, we are committed to work with industry to take forward the work identified as being required in the various sector strategies. To that end we attach real importance in the short term to:

- in the maritime sector, building on the close joint working that has been in hand for some months to develop a maritime industrial strategy. To this effect **we will immediately start negotiations with the key companies that make up the submarine supply chain** to achieve a programme level partnering agreement with a single industrial entity for the full lifecycle of the submarine flotilla, while addressing key affordability issues. The objective is to achieve this agreement in time for the award of the contract for the fourth and subsequent Astute class submarines. This will be matched by the implementation of a unified submarine programme management organisation within the MOD. **For surface ship design and build, we aim within the next six months to arrive at a common understanding of the core load required** to sustain the high-end design, systems engineering and combat systems integration skills that we have identified as being important. We expect industry to begin restructuring itself to improve its performance and shall build on the momentum generated by the industrial arrangements being put together on the CVF programme to drive restructuring to meet both the CVF peak and the reduced post-CVF demand. **For surface ship support, we will start immediate negotiations with the industry with the aim of exploring alternative contracting arrangements and the way ahead for contracting the next upkeep periods, which start in the autumn of next year.** Key maritime equipment industrial capabilities will be supported by the production of a sustainability strategy for these key equipments by June 2006;

- for fixed wing aircraft, developing the dialogue in which we have been engaged by **commencing negotiations with BAE Systems on the terms of the business rationalisation and transformation agreement required to facilitate the effective sustainment of the industrial skills, capability and technologies – wherever they may be in the supply chain – that will be so important to our ability to operate, support and upgrade our fast jet combat aircraft through-life.** We aim on working with the company during 2006 to agree the way ahead – which will be challenging given the scope of the scale of the transformation that is required – and to implement it from 2007. In parallel and contributing to these efforts, subject to value for money being demonstrated and appropriate commercial arrangements being put in place, we intend to move ahead with a substantial Uninhabited Aerial Vehicle Technology Demonstrator Programme in 2006;

- for Armoured Fighting Vehicles, working hard with BAE Systems, building on the discussions we have already set in train, and the agreement reached in December 2005, to **give effect to the long term partnering arrangement required to improve the reliability, availability and effectiveness through life of our existing AFV fleets.** Initial activity will focus on implementing measures that build confidence on both sides. We intend to establish a joint partnering team within the early part of 2006 and to establish a business transformation plan underpinned by a robust milestone and performance regime. The plan will detail the improvements in performance to be achieved, the process and behavioural changes required of both BAE Systems and the Department, and the capabilities and skills necessary to sustain through life support to AFVs;

- for helicopters, **driving forward with AgustaWestland the implementation of the business transformation partnering arrangement** to which we committed through the Heads of Agreement signed in April 2005. A partnering team (jointly resourced by MOD and AgustaWestland) has, for the last six months, been exploring partnering opportunities across those areas of the business indicated in the Heads of Agreement. **We hope that by the Spring 2006**, subject to value for money having been demonstrated, **we will have reached agreement on a Strategic Partnering Arrangement (SPA)** which will be focused on activities to sustain the design engineering skills and knowledge of UK military demands and safety standards within the company necessary for them to provide effective through-life support to those elements of the in-service helicopter fleet for which they are the design authority. The SPA is intended to commit both parties to specific targets on, inter alia, cost and schedule adherence (and where appropriate improvement) and improvements in operational availability; it will be underpinned by a Business Transformation Plan that sets out the process and behavioural changes required both in MOD and AgustaWestland;

- for complex weapons, establishing a multi-disciplinary team charged **with working with all elements of the onshore industry to establish how we might together seek both to meet our ongoing requirements and sustain in an industrially viable manner the critical guided weapons technologies and through life support capabilities that we judge to be so important to our operational sovereignty.** Given the transnational nature of the industrial players, this dialogue will need also to engage our allies and partners, particularly in Europe. This work will be complex and will necessarily take time, but our intention is that we should have a clearer way ahead by mid-2006;

- for general munitions, taking forward Project MASS, with a view to making **decisions on how best to sustain our required access to general munitions in the summer of next year,** building on the joint working arrangements enshrined in the existing Framework Partnering Agreement as reinforced by the recently agreed MOD/BAE Systems LS partnering principles. We are also actively pursuing partnering arrangements with other suppliers.

C2.5   In addition to this sector specific work, we will also work to give effect to the conclusions identified with respect to **science and technology.** Specifically, we will:

- **review the alignment of our research programme with MOD needs** and the needs of the defence industry, with a view to improving the alignment, quality and military exploitation of the research programme. This work is encompassed in the ongoing MOD Science and Technology Capability and Alignment Study and will be published by Summer 2006; it will be repeated every two years in concert with the biennial planning process;

- conduct **further work better to understand the underpinning technologies** that the UK must have for security and sovereignty reasons, where the UK is strong and where we need to focus our R&T efforts. We hope this will be completed by Autumn 2006;

- by mid-2006 **update our Defence Technology Strategy** to reflect the conclusions of the DIS and this related work. We will engage the R&T sub-group of the NDIC in this endeavour;

- develop a **better understanding of the innovation process** and map out the technology trees for major capabilities, systems and platforms in a report, which we aim to produce by the Autumn of 2006. We will work with the R&T sub-group of the NDIC to identify sources of technology and innovation throughout the supply chain and ensure that relevant technologies are pulled through into military capability.

C2.6   More broadly, we will place real effort and priority on **driving forward the programme of cultural, behavioural, procedural and where necessary organisational change** set out in Chapter C1. Our priorities for acquisition improvement are: partnering and relationships with industry, through-life; delivery of integrated solutions; agility and flexibility in projects; consistency in our approach; and professional delivery skills. Specifically, in the near term we:

- will work with industry to develop, roll out and implement a joint plan for **embedding the Defence Acquisition Values throughout the acquisition community. We expect to be in a position to launch this within three month**s and will apply the real commitment of resource, time and effort that will be required to effect lasting change through 2006 and beyond;

- are currently scoping **two Pathfinders programmes to test and de-risk a programme approach to through-life capability management** with the intention that the Pathfinder programme teams will launch in the first half of 2006;

- **will address the shortages in Project Delivery Skills** within the Department by building on our existing Project Delivery Skills Strategy to deliver an accelerated pan-Defence Project Delivery Skills programme that will identify and fill the critical gaps, in particular

in the areas of project and programme management and acquisition leadership. Key milestones are to be achieved by April 2006;

- will ensure that our **recruitment, reward and recognition** practices deliver acquisition staff of the quality we need in the numbers required whose behaviour demonstrates their commitment to our Defence Values for Acquisition. An evaluation of current incentivisation initiatives will be completed by October 2006;

- will **establish a strong and professional operating framework for how the Department conducts its acquisition business.** Under the DPA Forward programme, the DPA will be piloting the operating framework during 2006, working in conjunction with the other areas of acquisition to ensure consistency and alignment;

- will **review – so that we are in a position to make judgements about this by May 2006 – the extent to which the current process and organisational construct supports, encourages, hinders or obstructs the delivery of excellence** in acquisition. This would allow us to commit to changes that are required this side of the summer recess;

- are **looking forward to discussing further with industry – in the first instance through the commercial policy sub-group of the NDIC early in the New Year – our ideas about alternatives to competition** as a means where appropriate of assessing value for money, with a view to developing a concrete action plan for taking them forward;

- will start **with immediate effect, to deliver on our revised policy of providing industry with a better and longer term understanding of our future plans.**

C2.7   We will ke ep the progress of this work, and the extent to which real change is being demonstrated on the ground, under review within the MOD, through the Acquisition Policy Board reporting to the Minister for Defence Procurement. We will want formally to review progress with the NDIC regularly, and intend to offer the NDIC a detailed plan at its next meeting.

C2.8   This is an ambitious and demanding programme of concurrent activities aimed at delivering a step-change improvement in acquisition performance, underpinned by improved relationships between the MOD and its suppliers and enhanced confidence in our ability to sustain the core industrial capabilities, technologies and skills that are required to allow us effectively and in an appropriately sovereign manner, to operate our Armed Forces. We are committed to making it work and to investing the time, effort and resources to ensure that it does.

C2.9   We recognise that this will require tough decisions along the way; we shall not shirk them. We look to industry to rise to the challenge with us, recognising the opportunities for future prosperity that will ensue. The nation's Armed Forces, and indeed the nation's interests, require nothing less.

Acronyms

AACSTO	Aircraft and Aircrew CBRN Survive to Operate
AAF	Agile Air Force
ABRO	Army Base Repair Organisation
ABSV	Armoured Battlefield Support Vehicle
ACAD	Aircrew Chemical Agent Detector
AFV	Armoured Fighting Vehicles
AH	Attack Helicopter
ALARM	Air Launched Anti Radiation Missile
AMRAAM	Advanced Medium-Range Air-to-Air Missile
APDS	Armour Piercing Discarding Sabot
ARRC	Allied Rapid Reaction Corps
ASM	Anti Structures Munition
ASRAAM	Advanced Short-Range Air-to-Air Missile
ASTOR	Airborne Stand Off Radar
ASW	Anti-Submarine Warfare
ATGW	Anti Tank Guided Weapon
AWACS	Airborne Warning and Control System
BAES LS	BAES Land Systems
BDT 3	Biological Detection Tier 3
BERP	British Experimental Rotor Programme
BH	Battlefield Helicopters
BISA	Battlefield Information System Application
BRH	Battlefield Reconnaissance Helicopter
BSMC	Biological Surface Monitoring Capability
BVRAAM	Beyond Visual Range Air-to-Air Missile
BW	Biological Warfare
C2	Command and Control
C2IS	Command and Control Information Systems
C4ISR	Command, Control, Communication and Computing, Intelligence, Surveillance and Reconnaissance
C4ISTAR	Command, Control, Communication and Computers, Intelligence, Surveillance, Target Acquisition and Reconnaissance
CADMID	Concept, Assessment, Demonstration, Manufacture, In-Service, Disposal
CAP	Capability Area Plan
CASOM	Conventionally Armed Stand-Off Missile
CBM	Command and Battle Management
CBRN	Chemical, Biological, Radiological and Nuclear
CCII	Command, Control & Information Infrastructure
CCM	Counter Counter-Measures
CESG	Communications Electronics Security Group
CIS	Command Intelligence Systems
CMC	Chemical Monitoring Capability
COEIA	Combined Operational Effectiveness and Investment Appraisal
COMR	Civilian Owned Military Registered
CONDOR	Covert Night Day Operations Rotorcraft
COTS	Commercial Off the Shelf
CR2	Challenger 2
CSP	Capability Sustainment Programme
CT	Counter-Terrorism
CUP	Capability Upgrade Programme
CVF	Future Carrier Strike
CVR	Combat Vehicle Reconnaissance
CVRT	Combat Vehicle Reconnaissance Tracked
CWG	Capability Working Group
D&I	Detection and Identification
DA	Design Authority
DARA	Defence Aviation Repair Agency
DAS	Defensive Aid Systems
DATCCE	Deployable Air Traffic Control Capability Enhancement
DCPS	Deployable Collective Protection System
DDASC	DABINETT Development and Support Contractor
DEC	Director Equipment Capability
DEC(SP)	Director Equipment Capability (Special Projects)
DERA	Defence Evaluation and Research Agency
DESO	Defence Exports Services Organisation
DfID	Department for International Development
DFTS	Defence Fixed Telephone Service
DG	Director General
DHFCS	Defence HF Communications Service
DII	Defence Information Infrastructure
DIP	Defence Industrial Policy
DIS	Defence Industrial Strategy
DoD	Department of Defence
DS&S	Defence Strategy and Solutions
DSG	Defence Strategic Guidance
DTC	Defence Technology Centres
DTI	Department of Trade and Industry
DU	Depleated Uranium
ECC	Equipment Capability Customer
ECM	Electronic Counter Measures
EDA	European Defence Agency
EDEM	European Defence Equipment Market
EO	Electro Optic
EOCM	Electro Optic Counter-Measures
EP	Equipment Plan
EPW2	Enhanced Paveway 2
ESM	Electronic Support Measures
EW	Electronic Warfare
F&FS	Fuze and Fuze Setter
FACO	Final Assembly and Check Out
FAS	Future Army Structures
FAS GW	Future Anti-Surface (Guided Weapon)
FCO	Foreign and Commonwealth Office
FIAC	Fast Inshore Attack Craft
FIST	Future Integrated Soldier Technology
FMCMC	Future Mine Counter-Measures Capability
FPA	Framework Partnering Agreement
FRC	Future Rotorcraft Capability
FRES	Future Rapid Effects System
FSC	Future Surface Combatant
FSC	Field Standard C
FSTA	Future Strategic Tanker Aircraft
GCHQ	Governments Communications Headquaters
GMLRS	Guided Multi Launch Rocket System
GoCo	Government owned Contractor operated
GPS	Global Positioning System
GSR	General Service Respirator
GT	Gas Turbine
HE	High Explosive
HESH	High Explosive Squash Head
HF	High Frequency

HMI	Human Machine Interface	MVD	Medium Versatile Derivative
HMT	Her Majesty's Treasury	NBC R&S	NBC Reconnaissance & Survey
HUMS	Health and Usage Monitoring System	NBCD	Nuclear Biological Chemical Damage Control
HVM	High Velocity Missile	NDIC	National Defence Industries Council
IA	Improved Ammunition	NEC	Network Enabled Capability
IA	Integration Authority	NLAW	Next Generation Light Anti-Armour Weapon
ICT	Information Communication Technology	NSQEP	Nuclear Suitably Qualified and Experienced Personnel
IFPA	Indirect Fire Precision Attack	NSRP	Nuclear Steam Raising Plant
IFV	Infantry Fighting Vehicle	OA	Open Architectures
IM	Insensitive Munitions	OCCAR	Organisation Conjointe de Cooperation en Matiere d'Armament
IMS	Ion-Mobility Spectrometry		
INCOSE	International Council On Systems Engineering	OECD	Organisation for Economic Co-operation and Development
IOS	Integrated Operational Support	OFT	Office of Fair Trading
IP	Intellectual Property	OGDs	Other Government Departments
IP	Industrial Participation	OJEU	Official Journal of the European Union
IPR	Intellectual Property Rights	OSD	Out of Service Date
IPTs	Integrated Project Teams	OTS	Off the shelf
IR	Infra-Red	P3I	Pre-Planned Product Improvement
IS	Information Systems	PAAMS	Principal Anti Air Missile System
ISD	In-Service Date	P-BISA	Platform Battlefield Information Systems Application
ISMS	Integrated Sensor Management System	PBX	Polymer Bond Explosives
ITT	Invitation To Tender	PFI	Private Finance Initiative
J2CSP	Joint Command and Control Support Programme	PNPS	Personnel NBC Protection System
JAMES	Joint Asset Management and Engineering Solutions	PPP	Public Private Partnership
JCA	Joint Combat Aircraft	PSI	Prime Systems Integrator
JHC	Joint Helicopter Command	PWII	Paveway II
JMATS	Joint Military Air Traffic Services	PWIV	Paveway IV
JNIB	Joint Network Integration Body	R&D	Research and Development
JRRF	Joint Rapid Reaction Force	R&T	Research and Technology
JSF	Joint Strike Fighter	RAF	Royal Air Force
KERR	Kinetic Energy Risk Reduction	RCS	Radar Cross Section
LACP	Land Advanced Concept Phase	RDA	Regional Development Agency
LCAC	Landing Craft, Air Cushion	RDMS	Remote Delivered Mine System
LCAD	Lightweight Chemical Agent Detector	RDS	Rapid Diagnosis System
LEAPP	Land Environment Air Picture Provision	RF	Radio Frequency
LEP	Life Extension Programmes	RFA	Royal Fleet Auxiliary
LF ATGW	Light Forces Anti-Tank Guided Weapon	RM	Royal Marines
LFA	Low Frequency Active	RMS	Real-time Medical Surveillance System
LIMAWS	Lightweight Mobile Artillery Weapon System	RN	Royal Navy
LOVA	Low Vulnerability	SAA	Small Arms Ammunition
LPD	Landing Platform Dock	SALW	Small Arms and Light Weapons
LPH	Landing Platform Helicopter	SAM	Submarine Acquisition Modernisation
LRT	Light Role Team	SAR	Search and Rescue
LSD(A)	Landing Ship Dock(Auxiliary)	SAR	Synthetic Aperture Radar
LTPA	Long Term Partnering Agreement	SCMR	Surface Combatant Maritime Rotorcraft
MARS	Military Afloat Reach and Sustainability	SDR	Strategic Defence Review
MASC	Maritime Airborne Surveillance and Control	SDS	Surface Detection System
MASS	Munitions Acquisition – the Supply Solution	SEAD	Suppression of Enemy Air Defence
MBDS	Maritime Biological Detection System	SEMTA	Science, Engineering and Manufacturing Technologies Alliance
MCAD	Man-portable Chemical Agent Detectors		
MCM	Mine Counter Measures	SF	Special Forces
Med CM	Medical Countermeasures	SH-PRAC	Squash Head Practice Round
MEMS	Micro Electric Mechanical Machines	SLUW	Submarine Launched Underwater Weapon
MFTS	Military Flying Training School	SME	Small and Medium size Enterprises
MIS	Maritime Industrial Strategy	SOCD	Stand Off Chemical Detector
MJDI	Management of the Joint Deployed Inventory	SPA	Strategic Partnering Arrangement
MLD	Multi Level Decontamination	SPEAR	Selected Precision Effects at Range
MLI	Mid Life Improvement	SSBN	nuclear powered ballistic missile submarine
MMIT	Management of Materials in Transit	SSN	nuclear powered submarine
MODAF	Ministry of Defence Architectural Framework	SSS	Surface Ship Support
MOTS	Modified off the shelf	STDL	Secure Tactical Data Link
MTE	Military Task Equipment	STOVL	Short Take-Off and Vertical Landing
MTs	Military Tasks	STP	Short Term Plan

T&E	Test and Evaluation
TACAS	Tubed Artillery Conventional Ammunition System
TADL	Thales Air Defence Ltd
TBD	To Be Decided
TDP	Technology Demonstrator Programme
TEWA	Threat Evaluation and Weapon Allocation
TIH	Toxic Industrial Hazards
TLAM	Tomahawk Land Attack Missile
TRaME	Tactical Radiation Monitoring Equipment
TSB	Technology Strategy Board
TSCP	Transatlantic Secure Collaboration Programme
TSS	Transforming Submarine Support
TUM	Truck Utility Medium
UA	Unmasking Aid
UAV	Uninhabitated Air Vehicle
UBDS	Unmanned Biological Detection System
UCAV	Unmanned Combat Air Vehicle
UDS	Unit Decontamination Capability
UKASCACS	UK Air Surveillance Command and Control System
UKMFTS	UK Military Flying Training System
UN	United Nations
UOR	Urgent Operational Requirement
UxV	Unmanned x Vehicles (i.e. where x could be underwater, surface, air etc)
VMF	Versatile Maritime Force
VSC	Versatile Surface Combatant
WCSP	Warrior Capability Sustainment Programme
WLC	Whole Life Cost
WMD	Weapons of Mass Destruction